AF572312

Hazardous Materials Dictionary

Hazardous Materials Dictionary

RONNY J. COLEMAN
KARA HEWSON WILLIAMS

LANCASTER · BASEL

Published in the Western Hemisphere by
Technomic Publishing Company, Inc.
851 New Holland Avenue
Box 3535
Lancaster, Pennsylvania 17604 U.S.A.

Distributed in the Rest of the World by
Technomic Publishing AG

Printed in the United States of America
10 9 8 7 6 5 4 3

Main entry under title:
Hazardous Materials Dictionary

A Technomic Publishing Company book
Bibliography: p.

Library of Congress Card No. 87-71934
ISBN No. 87762-539-5

INTRODUCTION

Managing hazardous materials incidents is quickly becoming a necessary skill for industries and communities entrusted with public safety. Accidents involving dangerous substances continue to multiply. Thousands of persons have been injured and killed, or are destined to die prematurely, as a result of such mishaps. Names like *Kingman, Bhopal* and *Chernobyl* have become more than obscure places on the map. Their worldwide familiarity rests on their being sites where a hazardous materials incident has exploded into tragedy.

Hazardous substances are not going to disappear. They will increase, and we must become more aware of them and their impact on life, safety and the environment. The type and number of local, state and federal agencies created to attempt to control the problems have increased in direct proportion. Acronyms such as DOT, EPA, CHEMTREC and NTSB are familiar even to the layman. Moreover, the regulations, procedures and policies established by these agencies have proliferated rapidly.

Many communities have recognized this danger to their personnel and have instituted first-responder training, as well as hazardous materials response teams. Other agencies have formed sophisticated record-keeping systems to track down specific dangers from specialized chemicals.

The parallel growth of the chemical industry and of emergency response capabilities in the public and private sectors has created a new need for improved communication. A new vocabulary of important terms is emerging in each of the industries that transport, store and handle hazardous materials.

The precise definition of specialized words is critical for communication, especially under the stressful circumstances of an emergency. In such cases there is no margin for the error of using unclear terms, slang or buzzwords that can be subject to wrong interpretation. Professionals called upon to deal with hazardous materials incidents must be able to address one another in language that is readily understandable and conducive to appropriate action.

The field of hazardous materials management is somewhat unique in that it combines words from many different professions. One is tempted to think of hazardous materials terminology in terms of the materials only, but the field is much more complex than that. For example, a hazardous material contained in a railcar pressure vessel that is spilling or leaking into a waterway creates the needs to use words from several different disciplines.

The hazardous materials teams that must respond to these emergencies have to bridge the language difference between these various disciplines. As the training of first responders becomes more prevalent the terminology of these professions will also begin to merge with the jargon of the responding agencies.

The purpose of this book is to provide a reference source that identifies many of the unique terms that apply to the handling of hazardous materials emergencies. As far as we can determine this is the first attempt to accomplish this task. There are several sources for this material, but usually they have been limited in distribution to the specific occupation that generates the terminology. There have been listings of terms used by the trucking industry, railroad industry and environmental protection industries and so forth. However, there has not been a single source of these terms available.

This dictionary, representing a compilation of words and phrases from many relevant sources, will help document and standardize the nomenclature of hazardous materials. The authors have screened the technical discourse of the chemical, transportation, petroleum and medical fields, both governmental and private, to determine the most current expressions and their uses. The lexicographic goal has been to identify key terms, ambiguous and multiple meaning words, acronyms, symbols and even slang referring to hazardous materials reactions, storing and handling procedures.

The field of hazardous materials incident management is rapidly growing. It can be reasonably anticipated that this glossary will be added to substantially with each edition. All languages are dynamic and grow with practice, shrink with disuse or are modified by the use of common words to describe uncommon events or actions.

A

"A" End of Car (RR) The end opposite that on which the hand brake is mounted.

"A" Unit (RR) A diesel unit equipped with a cab and operating controls.

Abandon To cease efforts.

Abandoned Site An inactive hazardous waste disposal or storage facility which cannot be easily traced to a specific owner, or whose owner has gone bankrupt and consequently cannot afford the cost of cleanup, or a location where illegal dumping has taken place.

Abatement A method of reducing the degree or intensity of pollution such as the restoration, reclamation or recovery of natural resources adversely affected by that pollution; also, the use of such a method.

AB Brake (RR) The current standard freight car air brake system. *see* AUTOMATIC AIR BRAKE.

AB Control Valve The operating valve of the AB freight car air brake. Controls the charging, application and release of the brakes.

Abscissa The horizontal plane showing dosage values (or exposure time) with increasing magnitude from left to right on graphs representing toxicological mortality curves.

Absolute Block (RR) A block which a train is not permitted to enter while it is occupied by another train.

Absolute Filter A filter having a removal efficiency of at least 99.97% for .03 micron particles. Often called high efficiency particulate air filters (HEPA filters).

Absolute Pressure Gauge pressure plus atmospheric pressure.

Absorb To soak up or drink in, as when a hazardous material is taken into a plant, textile or soil.

Absorbed Dose The quantity of a material absorbed per unit of mass of tissue; unit of dose varies from chemical to chemical.

Absorbent Materials Any material used to soak up chemical materials. Examples: sand, sawdust, commercial bagged clay, kitty litter and zorbal.

Absorption (1) Movement of a chemical into a plant, animal or soil. Compare to ADSORPTION. (2) Any process by which one substance penetrates the interior of another substance. In oil spill cleanup, this process applies to the uptake of oil by capillaries within certain sorbent materials. *see* CAPILLARY ACTION.

Absorptive Clay A special type of clay powder which can soak up chemicals and hold them. This clay is often used to clean up chemical spills.

Acaricide A pesticide used to control spiders, ticks and mites; miticide.

Accelerant A chemical substance used to initiate or promote a fire. Flammable liquids are the most common types of accelerants.

Accelerator A device utilized in the field of nuclear engineering to increase the velocity and energy of charged elementary particles, e.g., electrons or protons, through the application of electrical or magnetic forces. Types of accelerators include betatrons, Cockcroft-Walton, cyclotrons, linear accelerators, synchrocyclotrons, synchrotrons, and Van De Graaff generators.

Access Road Any passage providing access to a treatment, storage, or disposal area within a Hazardous Waste Management (HWM) facility, suitable for use by transport vehicles and emergency vehicles in all types of weather.

Accident An uncontrolled event which has the potential for damaging life or property; synonym for INCIDENT.

Accident Mechanism The series of events which constitute the course of events culminating in the release of hazardous chemicals outside of their normal containment.

Accidental Explosion Unintentional detonation/ignition of explosive or suspected explosive materials not associated with criminal activity. Generally relates to some type of industrial or commercial activity.

Accumulate To build up, add to, store, pile up.

Accumulative Pesticides Those pesticides which tend to build up in the tissue of animals or remain persistent in the environment.

Acetylcholinesterase (AChE) In pesticides, an enzyme that will most rapidly hydrolyze acetylcholine as substrate, will not hydrolyze most non-choline esters, is inhibited by excess substrate, and is derived primarily from nervous tissue.

Acid A hydrogen-containing compound which reacts with water to produce hydrogen ions; a proton donor; a liquid compound with a pH less than or equal to 2. Acidic chemicals are corrosive.

Acid Suits Specialized protective clothing that prevents toxic or corrosive substances from coming into contact with the body of the wearer.

Acre 43,560 square feet. Commonly used term when assessing involved areas of spills and leaks. Equal in area to a plot 440 feet long by 99 feet wide; equal in area to a plot 209 feet by 209 feet; or a rectangle 436 by 100 feet.

Action What an ACTOR does or did during a particular event.

Activated Carbon A highly absorbent form of carbon, used to remove odors and toxic substances from gaseous emissions or liquid effluents.

Activator A material that is added to a pesticide to increase its toxicity.

Active Ingredients A term used in the development of pesticides. Refers to the chemical that has the toxic potential. Active ingredients are listed, in order, on a pesticide label as a percentage of weight or pounds per gallon of concentrate. To be contrasted to "inert" ingredients.

Active Portion A portion of a facility where hazardous waste treatment, storage, or disposal operations are being conducted subsequent to November 19, 1980 and which has not yet been closed.

Active Resources (ICS) Resources logged and assigned to work tasks during an emergency.

Activity (1) A set of actions directed towards an intended outcome or result. (2) Rate of decay of a radioactive sample. Measured in curies, i.e., number of nuclear transformations per second.

Act of God An unanticipated grave natural disaster or other natural phenomenon of an exceptional, inevitable and irresistible character, the effects of which could not have been prevented or avoided by the exercise of due care or foresight.

Actor A person who is involved in an activity and does something to effect an event.

Actual Dosage The amount of active ingredient (not a formulated product) of a pesticide that is applied to a specific area or target.

Acute Of short or intense course, not chronic; of recent or sudden onset seriously demanding urgent attention.

Acute Dermal Poisoning A single dose of toxic chemicals absorbed through the skin in amounts capable of causing death.

Acute Health Effects Health effects that occur or develop rapidly after exposure to a substance.

Acute Inhalation Poisoning A single dose of toxic chemicals absorbed into the lungs in amounts capable of causing death.

Acutely Hazardous Waste A waste considered to present a substantial hazard whether improperly managed or not. EPA includes in this category waste shown to be fatal to humans in low doses, those shown in mammalian studies to have specific toxicities, and explosives.

Acute Oral Poisoning A single dose of toxic chemicals absorbed into the digestive track in amounts capable of causing death.

Acute Poisoning A single exposure of a toxic substance of high toxicity that, if untreated, would be lethal.

Acute Radiation Exposure Exposure to high radiation levels over a short period of time, usually 24 hours or less.

Acute Radiation Syndrome Illness which develops as a result of acute exposure to high radiation levels; consists of cerebrovascular, gastrointestinal, and hematopoietic syndromes.

Acute Toxicity Short-term poisonous effects. The toxicity of a material determined at the end of 24 hours which causes death or injury from a single or limited exposure.

Additive Any substance added to a pesticide to improve performance. As as ADJUVANT.

Additive Effects Mixtures of insecticides in which the total toxicity amounts to the sum of the individual insecticides.

Adenosine Triphosphate (ATP) A compound present in living cells which provides energy derived from food or sunlight for the energy expending process (such as protein synthesis, muscle contraction, impulse conduction, and glandular secretion).

Adherence The characteristics of a material causing it to stick to another surface; a factor in contamination.

Adhesive An additive or adjuvant that helps a pesticide adhere to different surfaces.

Adhesion The attraction of unlike molecules. Molecular attraction which holds the surfaces of two substances in contact, such as water and rock particles.

Adiabatic Ignition A rapid compression of flammable vapors that generates a sufficient amount of heat to cause the ignition of those vapors. Synonomous with DIESELING.

Adjuncts (EMS) Equipment, special devices and drugs used by specially trained personnel to assist in performance of life support measures. Examples: airways, cardiac monitors, intravenous infusion, oxygen, lidocain.

Adjustable Undercarriage A truck undercarriage with a provision for convenient fore-and-aft adjustment of its location on the trailer; used to shift a greater part of a vehicle's gross weight or load onto the kingpin or suspension.

Adjuvant Any substance (usually an inert ingredient) added to a pesticide solution to make the formulation work better. Synonyms: adhesive, emulsifier, penetrant, spreader, wetting agent. Same as ADDITIVE.

Adsorb To collect, as a gas or liquid, in condensed form on the surface of another substance.

Adsorption (1) The process by which one substance is attracted to and adheres to the surface of another substance without actually penetrating its internal structure. (2) Process by which a substance is held (bound) to the surface of a soil particle or mineral in such a way that the substance is only available slowly. Clay and highly organic materials tend to adsorb pesticides rather than absorb them.

Adulterated Contaminated or made impure.

Advance of a Signal (RR) The side of the signal opposite to that from which the indication is received.

Advanced Life Support The administration of drugs, fluids, and other advanced medical care by paramedics in the pre-hospital setting.

Aerator A device used in the moving of dry bulk solids in order to improve the flowability of the material.

Aerobic Living in the air. Opposite of ANAEROBIC.

Aerosol A system in which liquid or solid particles are distributed in a finely divided state through a gas, usually air. Particles within aerosols are usually less than 1 micron (0.001 mm) in diameter, and are more uniformly distributed than in a spray.

Aerosol (Generator) A low concentration solution of a pesticide or combination of pesticides, usually in the form of oil solution especially formulated for use in generators.

Aerosol (Pressurized can) A small amount of material of any sort driven through a fine opening by an inactive gas under pressure. When nozzle is triggered it produces a fine spray, mist or fog of minute solid or liquid particles suspended in the air.

AFFF Aqueous Film Forming Foam; an extinguishing agent that may be used on many flammable liquids.

AGA American Gas Association.

Agent Freight agent.

Agitate To keep a chemical formulation mixed up; to keep a formulation from separating or settling out in a tank.

Agitation The act of mixing or stirring a chemical formulation.

Agitator The paddle or other mechanical device that uses air, hydraulic action or some other form of movement to keep a chemical formulation mixed in a vessel.

Agricultural Commodity Any plant, or part of a plant, animal or animal product that is to be bought or sold in commerce.

Agricultural Waste Waste materials such as rice straw, wood wastes, orchard prunings, bark, manure and cotton gin trash.

Agroecosystem The system of plants, animals and habitat used for human purposes.

Agronomist Scientist specializing in study of soil or plants as related to crop or crop production.

Aileron The movable, hinged portion of an airplane wing.

Air Bill Shipping papers prepared from bill of lading when hazardous materials are moved by an air carrier.

Air Blast Sprayer Machine used in pesticide applications to orchards, shade trees, vegetables and fly control.

Airborne Any substance transported by wind or wind currents.

Airborne Radioactives Any radioactive material dispersed in the atmosphere in form of dusts, fumes, vapors or gases.

Air Brake Hose (RR) Flexible connection between brake pipes of railroad cars or locomotives.

Air Compressor Power-driven air pump supplying compressed air for operation of air brakes and other air-actuated equipment.

Air Connection Fitting used to take air pressure from source to another pressure vessel; used on trucks, tanks, other cargo containers.

Air Gage (Gauge) Instrument indicating amount of air pressure in reservoirs or brake pipes.

Air Inversion Meteorological condition in earth's atmosphere in which the air some distance from earth surface is higher in temperature than that at ground level. Such condition traps air and released vapors near the earth surface, thereby impeding dispersion.

Air Lift Axle A single air-operated axle that will convert vehicle into multi-axle unit when lowered, providing vehicle with much greater load capacity; trucking industry term.

Air Pipe (Air Brake) *see* BRAKE PIPE.

Air Pollution Contamination of atmosphere with any material that can cause damage to life or property.

Air Pollution Control Districts/Air Quality Management Districts (APCDs/AQMDs) County and regional agencies established pursuant to the County Air Pollution Control Law of 1947 and the Mulford-Carrell Act of 1967 to administer and enforce minimum standards for air quality.

Air-Reactive (Materials) Substances that will ignite into open flame when exposed to open air.

Air Sampling Collection and analysis by instrument of samples of air to determine presence of hazardous materials.

Air Spring (Air bags) Flexible air-inflated chamber which controls air pressure and can be varied to support load and absorb road shocks. Can fail without warning, creating unstable load.

Air (or Water) Streams Method of oil containment where force of air or water directed as a stream can be used to divert or contain an oil slick.

Air Support Supervisor (ICS) Personnel responsible for development and implementation of air requirements of an emergency; reports directly to Operations Section Chief.

Air Tanker (ICS) Any fixed-wing aircraft certified by FAA as being capable of transport and delivery of fire retardant solutions.

Alcohols A class of organic chemical compounds containing hydroxyl group, used as solvents and in some preparations of chemical disperants. *see* GLYCOLS.

Alert A situation wherein apprehension exists as to the safety of an aircraft and its occupants.

Alert I Aircraft approaching an airport with minor difficulties.

Alert II Minor aircraft accident or incident or that the aircraft is experiencing major difficulties which may result in such an incident.

Alert III An aircraft has crashed on or near the airport, parked aircraft are endangered by fire or explosion, or aircraft is involved in a collision.

Alertor (RR) Device which detects frequency of engineman's movements and initiates an air brake application when required frequency is not maintained. *see* DEADMAN CONTROL.

Algicide A pesticide used to control algae, especially in stored water or industrial applications.

Aliphatic Chemical compounds comprised of straight chain molecules as opposed to a ring structure.

Alkali Any compound which forms the hydroxyl ion in its water solution. *see* BASE; synonyms include hydroxide, caustic.

Alkalinity The concentration of hydroxide ions.

Alkaloids Chemicals present in some plants, sometimes used as pesticide.

Alkanes A class of hydrocarbons which can be gas, solid, or liquids depending upon carbon content; its solids (paraffins) are a major constituent of natural gas and petroleum. Alkanes are usually gases at room temperature (methane) when containing less than 5 carbon atoms per molecule. Low carbon number alkanes produce anaesthesia and narcosis at low concentrations and at high concentrations can cause cell damage and death in a variety of organisms. Higher carbon number alkanes are not generally toxic but have been shown to interfere with normal metabolic processes.

Alkenes Another class of hydrocarbons also called olefins; sometimes are gases at room temperature but usually liquids; common in petroleum products. Generally more toxic than alkanes, less toxic than aromatics.

Alley (slang) (RR) A clear RR track for movement through a yard.

Alloy Mixture of two or more metals.

Alpha Particle A positive-charged particle emitted by certain radioactive materials. Consisting of two neutrons and two protons, is identical with nucleus of the helium atom. Least penetrating of the three common forms of radioactive substances (alpha, beta, gamma), it is not normally considered dangerous to plants, animals or people unless it gets into the body.

Alternative Technology The application of technology to reduction of waste generation, promotion of recycling and alternatives to land disposal of hazardous waste.

Alum Usually potassium aluminum sulfate used in water-treatment plants for settling out small particles of foreign matter; usually gelatinous when wet.

AMA Automobile Manufacturers Association.

Ambient Surrounding conditions, primarily used in reference to climatic conditions; ambient temperature.

Ambient Air Quality Standards Specified concentrations and durations of air pollutants reflecting relationship between the intensity and composition of air pollution to undesirable effects as established by a state board and/or the federal government.

Ambulance Service Area (EMS) Designated geographic area delineated by the local EMS agency to ensure availability of emergency medical transport services at all times by one or more specified providers.

Ammonia Stripping A steam process used in treatment of ammonia-bearing wastes. Process also used to remove volatile and organic contaminants from the waste stream. Ammonia readily condenses and can be reclaimed for use.

Anaerobic Able to live where there is no oxygen; opposite of AEROBIC.

Analeptics Medullary-stimulating drugs which increase respiration and blood pressure in mammals.

Analog A chemical compound similar in structure to another compound but differing in some slight structural detail.

Analysis The separation of a compound into its constituent parts.

Angle Cock (RR) A two-position valve located at both ends of the air brake on railroad cars. When open, allows passage of air.

Anhydrous Without water.

Aniline Point The lowest temperature a chemical aniline and a solvent (such as the oil in oil-base muds) will mix completely.

ANSI American National Standard Institute. Professional group cooperating to create voluntary standards.

Antagonism Decrease in effectiveness resulting from two or more chemicals being mixed together.

Antibiotic Substance used to control pest microorganisms.

Anticoagulant Chemical (usually a pesticide) preventing the normal clotting of blood.

Antidote Immediate treatment given to counteract effects of poisoning.

Antitranspirant A pesticide that reduces water loss by coating the leaves of a plant.

Aphicide Pesticide used to control aphids.

API American Petroleum Institute.

API Gravity A scale developed by API designating an oil's specific gravity or the ratio of weight of oil to pure water.

Application Directing a pesticide onto plants, animals, buildings, soils, water or any other area.

Applicator A person or piece of equipment which applies pesticides.

Approach Signal (RR) Fixed signal preceding an interlocking RR signal and governing the approach to the interlocking.

Apron Paved or hard-surfaced area to accommodate parked aircraft.

Apron Ring The lowest ring of plates in an oil tank.

Aqueous Indicates water is present in the solution.

Aqueous Treatment A hazardous waste treatment system designed to remove contamination from water so it may be safely returned to the environment.

Aquifer Underground zone of earth containing water, frequently is source for water supplies (wells); usually composed of gravel or porous material and can be contaminated by waste disposal or chemical releases.

Arbitrary and Capricious Without fair, solid and substantial reason.

Arenaceous Pertaining to sand or sandy rocks (such as shale).

Argillaceous A nonproductive formation consisting of clay or shale (such as argillaceous sand).

Aromatics Class of hydrocarbons considered to be the most immediately toxic; found in oil and petroleum products; soluble in water; considered long-term poisons and produce carcinogenic effects. Antonym: aliphatic.

Arrival Notice (RR) Notice furnished to consignee re:freight.

ARS Agricultural Research Service, division of USDA.

Arsenicals Pesticides containing the ingredient arsenic.

Arson Deliberate damage or destruction of property through use of an incendiary device, destructive device, or explosive that falls within the purview of ATF.

Aseptic Free of disease-causing organisms.

Ash The incombustible material remaining after a fuel or solid waste has been burned.

ASME American Society of Mechanical Engineers.

ASME Container Container constructed in compliance with ASME codes.

Asphalt Hydrocarbon material ranging in consistency from heavy liquid to a solid. Most common source is residue left after fractional distillation of crude oils; used primarily for surfacing roads.

Asphyxia Condition arising when the blood is deprived of adequate supply of oxygen; loss of consciousness or death may follow.

Asphyxiant Gas or vapor which, when inhaled, may lead to asphyxia.

Asphyxiating Materials Substances causing death by displacing of oxygen in the air. Example: carbon dioxide.

ASSE American Society of Safety Engineers.

Assigned Car (RR) A car that has been filled with a commodity and given a specific destination on the waybill or consists.

Assisting Agency (ICS) Any agency directly contributing suppression, rescue, support or service resources to another.

Associated Gas Natural gas, commonly known as gas-cap gas, which overlies and is in contact with crude oil in the reservoir.

ASTM American Society for Testing and Materials.

Asymptomatic Without symptoms.

ATA American Trucking Association.

ATF Alcohol, Tobacco, and Firearms division of the United States Department of the Treasury.

Atmospheric Pressure The pressure exerted over the surface of the earth by the weight of the atmosphere; at sea level, approximately 14.7 psi.

Atom A basic unit of physical matter indivisible by chemical means, the fundamental building block of chemical elements; composed of a nucleus of protons and neutrons, surrounded by electrons.

Atomic Cloud The hot gases, smoke, dust, and other matter carried aloft after an explosion of a nuclear weapon in the air or near the surface; frequently has a mushroom shape.

Atomic Energy Commission (AEC) The independent civilian agency of the federal government with statutory responsibility for atomic energy matters; the body of five persons appointed by the President to direct the agency.

Atomic Number Number of protons in nucleus of an atom. Each chemical element has been assigned a number in a complete series from 1 (hydrogen) to 103 (lawrencium).

Atomic Weapon An explosive weapon packaged for transportation or use by military forces in which the energy is produced by nuclear fission or fusion, as opposed to a nuclear device used for peaceful purposes such as energy production.

Atomic Weight The mass of an element relative to its atoms.

Atomize To break up a liquid into very fine droplets by forcing it through a nozzle-like device.

Atropine Antidote used by medical profession to treat poisoning by organic phosphates or carbamates.

Attack Line High-pressure hose used to apply water or foam to a fire, generally 1–1/2″ flowing 60–125 gpm.

Attapulgite A highly sorptive natural clay mineral having fibrous or spicular particles.

Attempted Bombing (Failure to detonate) Unsuccessful attempt to commit a bombing with high or low explosives or blasting agents, due to the malfunction, recovery or disarmament of an explosive device.

Attempted Incendiary Bombing (Failure to ignite) An unsuccessful attempt to commit an incendiary bombing due to the malfunction, recovery or disarmament of the device.

Attractant A chemical that lures insects, rodents or other pests to selected locations where they can be destroyed.

Autecology That branch of ecology dealing with the interrelationships of the individual organism and its environment.

Authority Governmental or legal body or a representative of such organizations having legal responsibility at a particular incident.

Authority to Construct Permit issued prior to construction of facilities which will emit a significant amount of pollutants to the atmosphere.

Authorized Registered Nurse (EMS) Personnel certified as qualified to provide emergency care and to issue emergency medical instructions to EMTs.

Authorized Representative Individual responsible for overall operation of a facility or a unit of that facility.

Auto-Ignition Temperature Minimum temperature to which a substance must be heated to initiate self-sustained combustion, independent of any open flame.

Automatic Aid A preplanned system where two or more fire departments respond routinely across jurisdictional boundaries to render mutual assistance; also: mutual aid.

Automatic Air Brakes (RR) A fail safe device that prevents a runaway train. It consists of an activation device that checks on the status of the train engineer. If the operator of the train is inactive for a specific period of time the brakes are automatically applied. Sometimes called a "deadman" switch or ATS (automatic train stop)

Automatic Control Device regulating various factors (as flow rate, pressure, or temperature) of a system without supervision or operation by personnel.

Autoradiograph A photographic record of radiation from radioactive material in an object, made by placing the object very close to a photographic film or emulsion; used to locate radioactive atoms or tracers in metallic or biological samples.

Auxiliary Power Unit Plant for generation of electrical power.

Available Resources (ICS) Personnel and material assigned to an incident, available for assignment.

Aviation Fuel Fuel designed to be used in commercial aricraft, including AVGAS, Jet A and Jet A-1, and Jet B types.

Avicide A pesticide used to control birds.

Avoid Prolonged Contact Phrase commonly used on pesticide labels warning against inhalation of vapors.

Axle A beam with spindles about which the wheels of a vehicle rotate.

Axle Camber The controlled convexity of a trailer axle allowing for deflection of the axle under the load, compensating for the curvature in the crown of road beds.

Axle Setting The distance from the centerline of a truck axle to the trailer rear surface.

Axle Weight Amount of weight transmitted to the ground by one axle or the combined weight of two axles in a tandem assembly.

Axon A nerve cell process usually single and elongate, terminating in short branches.

B

"B" End of the Car (RR) The end on which the hand brake is located.

"B" Unit (RR) A diesel unit without a cab and without complete operating controls. Usually equipped with hostler controls for independent operating at terminals.

Back End Loader Refuse truck which has power-driven loading equipment at the rear of the vehicle.

Background Radiation The radiation in our natural environment, including cosmic rays and radiation from the naturally radioactive elements, both inside and outside the bodies of people and animals; also called natural radiation.

Back Haul (RR) To haul a RR shipment back over part of the route which it has traveled.

Backpressure That pressure resulting from restriction of full natural flow of oil or gas.

Backscatter When radiation of any kind strikes matter (gas, liquid, or solid), some of it may be reflected or scattered back in the general direction of the source; important when beta particles are being counted.

Backshore Area of the shoreline above the high tide mark, only inundated with water during exceptionally high tides accompanied by high winds; does not support characteristic flora and fauna. Can supply granular materials for replacement of oil-contaminated beach material during shoreline cleanup programs.

Bacteria One-celled microorganisms.

Bactericide A pesticide used to control disease-causing organisms.

Bad Order RR car in need of repair.

Baffle An intermediate partial bulkhead in a tank truck which reduces the surge effect in a partially-loaded truck.

Bag In-Bag Out Method of removing items utilizing plastic bagging material from a contaminated enclosure to prevent spread of contamination.

Baler Machine used to compact waste materials to reduce volume, usually into rectangular bundles, for shipping, recycling or landfill.

Ballast (RR) Material placed on a RR roadbed to hold track in line.

Ballast Car A car used for carrying ballast material to repair site, usually a gondola or hopper.

Ballast Tamper A machine for compacting ballast under RR ties.

Barrel Unit of liquid measure used extensively by petroleum industry; equal to 35 Imperial gallons or 160 liters.

Barricade Shield A type of movable protection from radiation.

Barrier (Containment Barrier) Any non-floating structure constructed to contain or divert spilled material.

Base (1) Substance containing group forming hydroxide ions in water solution. (2) (ICS) A location where primary logistics functions are coordinated and administered during an emergency incident; an INCIDENT COMMAND POST.

Base Manager (ICS) Person responsible for activation of the Incident Command Post (Base), supervises base operations, distributes working materials for staff, provides food, sleeping and sanitation facilities for assigned personnel.

Base Hospital (EMS) One of a limited number of hospitals entering into a written contractual agreement with the local EMS agency responsible for directing the advanced life support system or limited advanced life support system assigned to it.

Base Soils Unconsolidated material (sand, silt, gravel, etc.) which separates the lower limits of refuse from groundwater and bedrock.

Basic Life Support Emergency care rendered by emergency medical technicians limited to stabilization of respiratory and circulatory functions and, if needed, the immobilization of the cervical spine. Does not include use of drugs or advanced telemetry equipment.

Basin Any uncovered device used to retain wastes as part of a treatment process, usually less than 100,000 gallons.

Beam A stream of particles or electromagnetic radiation going in a single direction.

Becquerel (BQ) The SI unit for measuring the activity of a radioactive sample.

Belt Line (RR) A short railroad operating within a city.

Bentonite A mica-like, highly water-absorbent derivative of volcanic ash that swells to several times its original volume, producing an alkaline pH because of sodium hydroxide release by hydrolysis.

Berm A ledge or shoulder, as along the edge of a paved road.

Beta A type of radiation, essentially an electron or positron, which can cause skin burns. Beta emitters are harmful if they enter the body but can be shielded by protective clothing.

Beta Particle An elementary particle emitted from a nucleus during radioactive decay.

BI- Prefix meaning two.

Bill of Lading Commercial document which accompanies a shipment of materials and lists all items in the shipment.

Binding Arbitration A process for the resolution of disputes wherein decisions are made by an impartial arbitrator; decisions of that arbitrator are binding.

Bioaccumulation Process occurring when toxic substances are passed through the food chain from soil to plants to grazing animals to human beings.

Bioassay The utilization of living organisms to determine the biological effect of some substance, factor or condition.

Biochemical Oxygen Demand (BOD) The amount of oxygen required by bacteria to stabilize decomposable organic matter under aerobic conditions.

Bioconcentration Process of a pesticide being concentrated in the tissues of plants or animals.

Biodegradable Waste material capable of being broken down by bacteria into basic elements.

Biodegradation The breaking down of substances into basic elements.

Biological Agent Microorganisms (primarily bacteria) added to the water column or soil to increase the rate of biodegradation of spilled oil. Also, nutrients added to water to increase growth.

Biological Control The use of living organisms to control pests; opposite of using pesticides.

Biological Hazardous Wastes Any substance of human or animal origin other than food wastes, which is to be disposed of and could harbor or transmit pathogenic organisms; includes but not limited to pathological specimens such as tissues, blood elements, excreta, secretions, bandages, and related substances.

Biological Magnification The concentration of certain substances in a food chain; an important mechanism in the concentration of pesticides and heavy metals in organisms such as fish.

Biological Productivity The rate at which the energy of the sun is transferred to and reflected in growth and/or abundance of plants and animals; an important consideration in assigning priorities for hazardous materials spill cleanup.

Biological Treatment Process by which hazardous waste is rendered non-hazardous or reduced in volume by the actions of microorganisms. The five principal techniques include activated sludge, aerated lagoon, trickling filters, waste stabilization ponds, and anaerobic digestion.

Biomagnification The increase of concentration of a pesticide in tissues of plants or animals.

Biomedical Telemetry (EMS) Transmission of biological data from a patient to a monitoring point by means of radio or wired circuits.

Bipyridyls A group of synthetic organic pesticides which includes the herbicide Paraquat.

Blasting Agents Any material or mixture consisting of fuel and oxidizer intended for blasting, not otherwise defined as an explosive.

Blast Wave A pulse of air, propagated from an explosion in which the pressure increases sharply at the front of a moving air mass, accompanied by strong, transient winds; shock wave.

Bleed Valve (Bleeder) A valve placed to drain off liquid or gas slowly.

BLEVE Abbreviation for Boiling Liquid Expanding Vapor Explosion. Normally associated with fires that involve compressed gases in cylinders; rapid rupture of a vessel caused by over-pressure accompanied by rapid burning of the tank contents.

Blow-Down Valve Manually operated valve on a cargo tank that quickly reduces the tank pressure to the atmospheric pressure.

Board (RR) (1) A fixed signal regulating railroad traffic and usually referred to as a Slow Board, Order Board, Clear Board or Red Board. (2) A list of employees available for service, also called Extra Board.

Boarding Car (RR) A car used as a place of lodging for workmen.

Body Center Plate The center plate attached to the underside of a truck body holster; *see* CENTER PLATE.

Bogie (1) The running gear of a highway semi-trailer which may be removed. (2) Term used to refer to a swivel railway truck. (3) On aircraft, a tandem arrangement of landing gear wheels that swivel up and down so all wheels follow the ground as the altitude of the plane changes or as ground surface changes.

Boiling Point The temperature when a substance changes from a liquid to a gas.

Boiling Water Reactor A nuclear reactor in which water, used as both a coolant and moderator, is allowed to boil in the core; resulting steam is used to drive a turbine.

Bolster Cross member on underside of truck body through which the weight is transmitted.

Boom (1) Containment boom—A floating mechanical structure which extends above and below the water surface, designed to stop or divert the movement of an oil slick. (2) In pesticides a section of pipe or tubing connecting several nozzles to apply a pesticide over a wider area.

Boom Failure Failure of a containment boom due to excessive winds, waves or currents or improper deployment.

Botanicals Pesticides made from plants.

Box Car An enclosed RR car used for general service, especially for cargo which must be protected from the weather.

Boyle's Law A chemical law relating to the condition of gases. PV = Constant; volume of a given mass of gas varies inversely with the absolute pressure if temperature remains constant.

Brake Club Stick used by freight trainmen to tighten hand brakes.

Brake Cylinder (Air brakes) Cylinder containing a piston forced outwardly by compressed air to operate brakes. When air pressure is released, piston is returned to its normal position by release spring coiled about the piston rod.

Brakeman RR employee who assists with train and yard operations.

Brake Pawl (Hand brake) A small, specially-shaped steel piece pivoted to engage teeth of brake ratchet wheel to prevent turning backward, thus releasing brakes.

Brake Pipe Air brake piping of a locomotive or RR car acting as a supply pipe for the reservoirs.

Brake Ratchet (Hand brake) Wheel attached to the brake shaft with teeth which the pawl engages.

Brake Shaft A shaft on which a chain is wound and by which the power of a hand brake is applied to the wheels.

Brake Shoe Friction material shaped to fit the tread of the wheel when the brakes are applied.

Brake Valve (Air brake) The valve by which the engineer operates the brakes.

Brake Wheel An iron wheel attached to upper end of the brake shaft which is manually turned to apply the brakes.

Brand Name Name given to a chemical product by the manufacturer. May or may not reflect the characteristics of the material.

Breathing Apparatus A self-contained, compressed air, pressure/demand (positive pressure) respiratory device designed to provide full respiratory protection; seals the airway against contamination.

Breeder Reactor A nuclear reactor producing fissionable fuel as well as consuming it, especially any reactor that produces more fuel than it consumes.

Bremsstrahlung Electromagnetic radiation emitted when a fast moving charged particle loses energy upon being accelerated and deflected by the electric field surrounding a positively charged atomic nucleus. S-rays produced in an ordinary X-ray machine are bremsstrahlung.

Brine Water saturated with salt.

Brine Channels Small passages formed in the lower surface of first-year sea ice formed by the exclusion of saline water during rapid freezing.

BTU British Thermal Unit, a measuring unit of heat.

Brush Patrol Unit (ICS) A light mobile unit having limited pumping water capacity, capable of off-road operations.

Bubble Barrier A containment barrier used with some success in calm harbors for oil spills.

Bubble Chamber A device used for detection and study of elementary particles and nuclear reactions.

Buffer Zone Minimum acceptable space between active portion of a hazardous waste facility and the property line. Usually under control of state or local standards.

Buildup Accumulation of a hazardous chemical; usually applied to pesticides.

Bulk Container A cargo container, such as may be transported by truck or railroad, or an ocean-going vessel designed to transport large quantities of a single product.

Bulk Freight Freight not in packages or containers.

Bulkhead A structure inside a cargo tank to prevent cargo from shifting or separating during transit.

Bulkhead Flat A flat RR car with adjustable bulkheads at each end of car, used for plywood, wallboard, etc.

Bulldozer Company (ICS) Any bulldozer with a minimum complement of two persons.

Bulletin Updates on the latest development.

Bumping Post (RR) A braced post placed at end of a stub track to prevent rolling cars from going off ends of the rails.

Bunching The accumulation and tender of RR cars for loading or unloading in excess of orders or contrary to customary schedules.

Bung A cap of screw-type device to close the small opening in the top of a metal drum or barrel.

Bunker "B" A relatively viscous fuel oil (No. 5 fuel) used primarily as a fuel for marine and industrial boilers.

Bunker "C" A very viscous fuel oil (No. 6) used as a fuel for marine and industrial boilers.

Bureau of Explosives An agency of the American Association of Railroads.

Bureau of Solid Waste Management The branch of DER responsible for planning, directing, evaluating, coordinating and organizing a statewide waste management and enforcement program including the Hazardous Waste Management Program.

Burning Agent Compounds such as gasoline which are used to ignite and sustain combustion of material that would not otherwise burn.

Burning Point (Fire point) That temperature when a substance sustains combustion.

By-Product A material produced in addition to the main product during the manufacturing process. May be a waste or commercial product.

C

C Centigrade degrees, Celsius.

Cab The space in a locomotive "A" unit containing the operating controls and providing shelter and seats for the engine crew.

Cab Signal (RR) A device located in the cab of a locomotive indicating condition of the track ahead by a display of signals.

Cab-Behind Engine Motor truck or truck tractor with driver's compartment and controls located at the rear of a hood-enclosed power plant.

Cab-Over Engine Motor truck or truck tractor with a substantial part of its engine located under the cab.

Cabin Car Caboose.

Caboose A RR car attached to rear of freight train for office use and living accommodations while in transit.

Caboose Valve (RR) A valve located in the caboose so emergency brake applications may be made from rear end of train.

Calcareous Deposit or Growth A growth of calcium carbonate, calcium, or lime on a surface.

Calibrate To measure amount of pesticide that will be applied by the equipment to the target in a given amount of time.

Calibrated Measured.

Calibration (1) A procedure to ensure the accuracy of instrument measurements. (2) Measurement of the delivery rate of application equipment.

California Air Resources Board (CARB) A statewide agency, established by the Mulford-Carrell Act of 1967, responsible for coordinating efforts to attain and maintain ambient air quality standards, for conducting research into the causes of and solutions for air pollution, and for regulating motor vehicles to reduce air pollution.

California Environmental Quality Act (CEQA) Enacted in 1970, this state legislation provides a process requiring governmental decision-makers to consider the environmental effects of their decisions and to take feasible measures to prevent significant, avoidable damage to the environment.

Caller A RR employee who notifies employees to report for duty.

Calorie The amount of heat required to raise the temperature of one gram of water one degree centigrade.

Camp (ICS) Geographical site within the general incident area but separate from the base, equipped and staffed to provide food, water, rest and sanitation services to incident personnel.

Cancelled A pesticide use that is no longer registered as a legal use by the EPA; remaining stocks can only be used by order of the Administrator, EPA. (Note: This order is less severe than SUSPENDED.)

Cancer A group of some 200 diseases grouped together because of their similar growth processes. Each cancer is believed to originate from a single "transformed" cell which does not respond to normal controls over growth, regardless of the part of the body it affects.

Canister A metal or plastic container filled with absorbent materials which filter fumes and vapors from the air before they are breathed in by an applicator.

Capacity Measure of quantity.

Capacity Indicators Device installed on tank to indicate capacity at specific level; marker.

Capillary Action Absorption.

Captive Facilities Facilities located on lands owned by a generator of hazardous waste and operated to provide for the treatment or disposal solely of that generator's hazardous waste. *see* ON-SITE FACILITY.

Car Distributor RR employee assigned responsibility of distributing empty freight cars.

Car Dumper (RR) Device for quick unloading of bulk materials such as coal or grain. After being clamped to the rail, the car is tilted or rolled over to discharge lading.

Car Float Large flat-bottomed boat equipped with tracks on which RR cars are moved in inland waterways.

Car Knocker (slang) RR car inspector.

Car Lining Material placed on walls of RR car for protection of cargo.

Car Mile Movement of a RR car the distance of one mile. Term is used in statistical data.

Car Seals *see* SEALS.

Car Stop A device for stopping motion of a RR car by engaging the wheels, as distinguished from a bumping post which arrests motion upon contact with the coupler of a car. *see* BUMPING POST.

Carbamate A synthetic organic pesticide which contains carbon, hydrogen, nitrogen, and sulpher; belongs to a group of chemicals which are salts or esters of carbonic acid. Carbamates may be fungicides, herbicides, or insecticides. Examples: aldicarb, carbaryl, carbofuran, and methomyl.

Carbon Dioxide A colorless, odorless inert gas that is a by-product of combustion.

Carbon Monoxide A highly toxic and flammable gas that is a by-product of incomplete combustion. Very dangerous even in very low concentrations.

Carbon sorption Process whereby activated carbon, known as the sorbent, is used to remove certain wastes from water by preferentially holding them to the carbon surface.

Carbonaceous Compound having a carbon atom in the chemical structure.

Carcinogen A substance or agent capable of producing cancer.

Carcinogenic Term describing the cancer-producing property of a substance or agent; capable of causing cancer.

Card Rack (RR) A small receptacle on the outside of a freight car to receive cards giving shipping directions.

Cardboard A small board secured to the outside of freight car containing cards giving shipping directions or warning of dangerous lading, etc.

Cardiopulmonary Resuscitation (CPR) (EMS) Opening and maintaining an airway, providing artificial ventilation, and providing artificial circulation by means of external cardiac compression as defined by the American Health Association.

Cargo Tank (Primarily a DOT designation) A container used to transport LP-Gas over the highway as liquid cargo, either mounted on a conventional truck chassis or as an integral part of a transporting vehicle in which the container constitutes in whole or in part the stress member used as a frame. Essentially a permanent part of the transporting vehicle.

Carload Quantity of freight required for the application of a carload rate for setting transportation fees.

Carrier Rig Self-propelled, wheeled unit used to service oil and gas wells, usually having the mast, hoist, engines, and other auxiliaries needed for service.

Cartridge The part of the respirator which absorbs fumes and vapors from the air before the applicator breathes them in.

Casinghead Gas (Oil well gas) Dissolved gas produced along with crude oil from oil completions.

Casing Pressure Gas pressure built up between the casing and tubing.

Casualty Collection Point (CCP) (EMS) Site for the congregation, TRIAGE (sorting), preliminary treatment, and evacuation of casualties following a disaster.

Catalyst A substance which, when present even in small quantities, affects the rate of a chemical reaction but is unchanged itself by the reaction.

Catchment Area (EMS) Geographic area served by a specified health care facility or EMS agency.

Catenary A system of wires suspended between poles and bridges supporting overhead contact wires normally energized at 11,000 volts.

Causal Organism The organism that produces a specific disease.

"Caution" A signal word used on pesticide labels to denote slightly toxic pesticides (Toxicity Categories III and IV) as defined by FIFRA (Amended).

CDFA California Department of Food and Agriculture.

Cell The basic biological unit of plant and animal matter.

Celsius Temperature scale used by scientists on which water freezes at 0° and boils at 100° at one atmosphere pressure.

Center Dump Car A RR car which will discharge its entire load between the rails.

Center Plate One of a pair of plates which fit one into the other and which support the car body on trucks, allowing them to turn freely under the car. The center pin or king bolt passes through both, but does not really serve as a pivot. The male or body center plate is attached to the underside of the body bolster. The female or truck center plate is attached to the topside of the truck bolster.

Center Sill The central longitudinal member of the underframe of a RR car which forms the backbone of the underframe and transmits most of the buffing shocks from one end of the car to the other. Freight cars with cushioned underframes use a special type of floating center sill construction.

Centigrade Celsius.

Centrifugation A hazardous waste physical treatment process by which heavier particles in the fluid move to the walls of a rotating vessel and are removed.

Cerebrovascular Syndrome Illness caused by exposure to extremely high doses of ionizing radiation, resulting in severe damage to the central nervous system.

Cerenkov Radiation An eerie blue glow given off by electrons traveling in a transparent material such as water; visible during operation of some nuclear reactors.

Certification The recognition by a certifying agency that a person is competent and thus authorized to use or supervise the use of restricted-use pesticides.

Certification Label A label permanently affixed to forward left side of the trailer stating that the vehicle conforms with all applicable Federal Motor Vehicle Safety Standards in effect on the date of original manufacture.

Certified Applicator An individual who is certified to use or supervise the use of any restricted use pesticides covered by his/her certification.

CFR Code of Federal Regulations.

CGA Compressed Gas Association.

Chain Reaction A reaction that stimulates its own repetition.

Charging *see* FILLING.

Charles' Law V/T = Constant; the volume of a given mass of gas is directly proportionate to the absolute temperature if the pressure is kept constant.

Chemical Barrier Chemicals acting as surface tension modifiers to inhibit the spread of an oil slick on water; short-lived.

Chemical Dispersion In oil spills, the process of spraying chemical dispersants to remove stranded oil from areas not considered considered biologically sensitive.

Chemical Name Scientific name designating the contents or formulae of the active ingredients of the pesticide.

Chemical Oxygen Demand (COD) A means of measuring the pollution strength of domestic and industrial wastes based upon the fact that all organic compounds, with few exceptions, can be oxidized by the action of strong oxidizing agents under acid conditions to carbon dioxide and water.

Chemical Process A particular method of manufacturing, usually involving a number of steps or operations.

Chemical Properties Properties of a material that relate to toxicity, flammability, or chemical reactivity.

Chemical Specialties Manufacturers Association (CSMA) An organization formed to set standards for insecticidal performance against standard test organisms and procedures.

Chemical Sterilants (Chemosterilants) A chemical that can prevent reproduction in insects (causes sterility).

Chemical Treatment The processes by which hazardous waste is rendered less hazardous or suitable for transport by changing the chemical composition of such waste. Principal techniques include neutralization, precipitation, ion exchange, chemical dechlorination, and chemical oxidation/reduction.

CHEMTREC Chemical Transportation Emergency Center.

Chlorates Substances which can be herbicides and defoliants; they act as contact poisons, are translocated, and may be absorbed from the soil to kill both plant roots and tops.

Chlorinated A synthetic organic pesticide that contains chlorine, carbon, and hydrogen; examples: DDT, endrin, lindane.

Chlorinated Hydrocarbon Chemical in which carbon, chlorine and hydrogen are inseparably combined.

Chlorolysis Hazardous waste chemical treatment method recycling chlorinated organic compounds and converting them into useful industrial products through addition of chlorine.

Chlorosis The yellowing of a plant's normally green tissue because of a partial failure of the chlorophyll to develop.

Chocolate Mousse A water-in-oil emulsion containing 50–80% water.

Cholinesterase A body enzyme (chemical catalyst) found in animals and humans that helps regulate the activity of nerve impulses and is necessary for proper nerve function. It is destroyed or damaged when organic phosphates or carbamates enter the body in any fashion.

Cholinesterase Inhibitor Any carbamate, organophosphate, or other pesticide that can interrupt the action of enzymes.

Chronic Health Effects Long-term effects from a one-time or a repeated exposure to a substance.

Chronic Poisoning Poisoning which occurs as a result of repeated exposures over a period of time.

Chronic Toxicity Small, repeated doses of a poisonous substance absorbed or ingested by animal or person over a period of time.

Circuitous Route An extremely indirect route.

Circus Loading (RR) Method of loading highway trailers by moving them over the ends of the cars.

Clarifier A gravity apparatus for removing settled solids from a fluid.

Class 1 Disposal Sites Areas where complete protection is provided for all time for the quality of ground and surface waters from all wastes deposited therein, and where protection against hazard to public health and wildlife resources is provided.

Classification To arrange or sort into uniform categories or classes, usually by size, weight, color, organic/inorganic, etc. (RR freight cars)—A destination and routing code used on switch lists for ease in switching cars.

Classification of Explosives

CLASS A EXPLOSIVE—A material or device presenting a maximum hazard through detonation.

CLASS B EXPLOSIVE—A material or device presenting a flammable hazard and functions by deflagration.

CLASS C EXPLOSIVE—A material or device containing a restricted quantity of either Class A or Class B explosives or both, but presents a minimum hazard.

Classification of Fires

CLASS A—Fires in combustibles and flammable gases or liquids.

CLASS B—Fires in combustible solids, such as wood, paper, cardboard.

CLASS C—Fires involving energized electrical equipment.

CLASS D—Fires involving combustible metals such as magnesium, zirconium, etc.

Classification—Pesticides The process of assigning pesticides into groups according to common characteristics; may be by target species, by chemical nature, manner of formulation, mode of action or toxicity.

Classification—Poisons

CLASS A POISON—A poisonous gas or liquid of such nature that a very small amount of the product is dangerous to life.

CLASS B POISON—U.S. Dept. of Transportation category for highly toxic material, a substance known to be so toxic to human life that it affords a severe health hazard during transportation.

Classification Yard A yard where RR cars are grouped according to their destinations and made ready for proper train movement.

Class Rate A rate based on an assigned class rating (a percentage of first class) published in the Uniform Freight Classiciation.

Clean Room A room designed to maintain a defined level of air cleanness under operating conditions.

Cleanout Fitting A fitting installed in the top of a tank to facilitate washing of the tank interior.

Clear Board (RR) A signal indication displayed to advise that no train orders are being held.

Clearance Lamps Lamps which show to the front or rear of a vehicle mounted on the permanent structure of the vehicle to indicate overall width of the vehicle, one on each side of the vertical centerline, at the same height, and as near the top as practicable.

Clearance or Clearance Limit (RR) The limiting dimensions of height and width for cars in order that they may safely clear all bridges, tunnels, station platforms and other structures as well as equipment on adjacent tracks.

Cleat A strip of wood or metal used to afford additional strength, to prevent warping, or to hold in position.

Clinical Pertaining to the symptoms and causes of disease.

Closed Portion Any portion of a facility where hazardous waste treatment, storage, or disposal operations have been closed.

Closure Actions taken by owner or operator of a hazardous waste facility to prepare the site for long-term care and to make it suitable for other uses after wastes are no longer accepted.

CNS Central Nervous System.

Coal Car (RR) A car for carrying coal, usually a hopper car.

Cobalt Bomb Theoretically, a nuclear weapon encased in cobalt, which would produce large amounts of the highly-penetrating and long-lasting gamma radiation emitted by cobalt 60.

Code of Federal Regulations (CFR) The formal name given to those books or documents containing specific regulations provided for by the law.

COFC (RR) Container-on-flatcar.

Coffin A thick-walled container (usually lead) used for transporting radioactive materials.

Cohesion The attraction of like molecules.

Coke Rack (RR) A slatted frame or box applied above the sides and ends of gondola or hopper cars to increase the cubic capacity for the purpose of carrying coke or other freight in which the bulk is large relative to the weight.

Collection The act of picking up waste material at homes, businesses, or industrial sites, and hauling it to a facility for further processing, transfer to large vehicles or disposal.

Collection Center A facility designed to accept materials from individuals, usually for recycling.

Combination Rates A rate made by combining two or more rates published in different tariffs.

Combination Stop and Tail Lamps An electrical device having the dual purpose of a stop lamp and tail lamp.

Combustibility or Combustible A material that can, under most conditions, enter into the combustion process; burnable.

Combustible Liquids Liquids with a flash point above 140°F according to NFPA definition.

Combustible Metals Metal which will burn.

Combustible Solids Those materials which ignite relatively easily and are subject to rapid flame propagation at moderate temperature.

Combustion Fire. An exothermic chemical reaction due to rapid oxidation of a fuel, involving light and heat.

Combustion Explosion Sudden fracture of a container (structure) accompanied by a shock wave (sound) due to overpressure created by the expansion of a gas (often mainly air) within the container due to absorption of heat produced by combustion of a flammable mixture within the structure.

Combustion Product Material generated during the burning or oxidation of matter.

Command (ICS) The act of directing, ordering and/or controlling resources (personnel and/or material) by virtue of explicitly legal or delegated authority.

Command Post Central control point for an incident, located at a safe distance from an accident site where the coordinator, responders and technical representatives can make response decisions, deploy manpower and equipment, maintain liaison with media, and handle communications.

Command Staff (ICS) Personnel who report directly to Incident Commander, usually Information Officer, Safety Officer, and Liaison Officer.

Commercial Waste Waste material originating in wholesale, retail, or service establishments such as office buildings, stores, hotels, universities, and warehouses.

Commodity Rate A rate applicable to a specific commodity between certain specified points.

Common Exposure Route A likely way (oral, dermal, respiratory) that a pesticide may reach and/or enter an organism.

Common Name A well-known, simple name of a pesticide accepted by the Pesticide Regulation Division of the EPA.

Common Name of Pesticide Well-known made-up name accepted by the Environmental Protection Agency to identify the active ingredients in a pesticide.

Communications Center That location within a general incident area equipped and staffed to provide and control communications for an emergency situation.

Communications Officer Individual on Command Post staff responsible for flow of information and maintenance of the communication system at the incident.

Communications System A collection of individual communication networks, transmissions systems, relay stations, control and base stations capable of interconnection and interoperations designed to form an integral whole. The individual components must be technically compatible, employ common procedures and operate in unison.

Compactor Any power-driven mechanical equipment designed to compress and reduce the volume of waste materials.

Compactor Truck A large truck with enclosed body having special power-driven equipment for loading and compressing waste materials.

Compartment Tank Cars A tank car with the body divided into several sections for the purpose of carrying different commodities in each compartment or smaller shipments.

Compartmentizer Car A box car equipped with movable bulkheads which can be used to divide the car into separate compartments.

Compatibility The mixture of two or more pesticides without their effectiveness being reduced or altered.

Compatible Capable of mixing two or more compounds without affecting properties of the other.

Compensation Payments awarded either through the courts or a government-administered fund to cover injury or damage caused by exposure to hazardous substances. Awards usually cover lost income, out-of-pocket medical expenses, and pain and suffering.

Competency-Based Curriculum (EMS) A program wherein specific objectives are defined for each of the separate skills and successful completion of an examination demonstrating mastery of such skills.

Complexing Agent A material which forms a chemical complex with a second material (very tightly bound at the molecular level) when the two come in contact.

Compost Decomposed organic matter, often used for soil enrichment.

Composting Controlled method of decomposing organic matter by the natural activity of microorganisms.

Compound A substance consisting of two or more elements that have been united chemically.

Compressed Gas Any material or mixture having in the container an absolute pressure exceeding 40 psia at 70°F or regardless of the pressure at 70°F, having an absolute pressure exceeding 104 psia at 130°F.

Compressed Gas in Solution A non-liquefied gas that is dissolved in a solvent, but at high pressures.

Compton Effect The glancing collision of a gamma ray with an electron. The gamma ray gives up part of its energy to the electron.

Concealed Damage Damage to the contents of a package which is in good order externally.

Concentrate A pesticide formulation (liquid or dry) as it is sold before diluting; usually contains a high percentage of active ingredient to save shipping and storage charges and yet is of convenient strength and composition for dilution.

Concentration Amount of active ingredient contained in a given volume or weight.

Concentration Limits The level for each hazardous waste constituent which triggers initiation of a corrective action program.

Condensate Hydrocarbons which are in the gaseous state under reservoir conditions but which become liquid either in passage up the hole or at the surface.

Condition A particular state of being of a person or thing or set of circumstances.

Conditional Use Permit (CUP) A discretionary permit issued by cities and counties and required for certain projects to impose conditions designed to assure that the project is compatible with the local General Plan and zoning ordinances and that impacts to neighboring land uses are minimized.

Conduction A method of heat transfer whereby molecules of the substance transfer heat from one to another by direct contact.

Conductor Train service employee in charge of train or yard crew.

Conflicting Routes Two or more routes over which simultaneous movements of cars cannot be made without possibility of collision.

Connate Water Water inherent to the oil-producing formulation or fossil sea water trapped in the pore spaces of sediments during their deposition.

Connecting Carrier A railroad which has a direct physical connection with another or forms a connecting link between two or more railroads.

Connection Box Contains fittings for trailer emergency and service brake connections and electrical connector to which the lines from the towing vehicle may be connected. Also: junction box, light box, bird box.

Consignee The person who is to receive a shipment.

Consignee Marks A symbol placed on packages for export, generally consisting of a square, triangle, diamond, circle, cross, etc. with designed letters and/or numbers for the purpose of identification.

Consignor Person or firm from whom shipment originates. Also called shipper.

Consist A rail shipping paper similar to a cargo manifest. May contain a list of cars in the train in order or a list of those cars carrying hazardous materials and their location on the train.

Constructive Placement When, due to some disability on the part of the consignor or consignee, a RR car cannot be placed for loading or unloading, it is considered to be under Constructive Placement and subject to demurrage rules and charges, the same if it were actually placed.

Contact Herbicide A pesticide that kills primarily by contact with plant tissue rather than as a result of translocation. Only that portion of a plant contacted is directly affected.

Contact Poison A pesticide which kills when it touches or is touched by a pest.

Container Any portable device in which material can be stored, handled, transported, treated, or disposed of.

Container (Explosives) The type of article/material/substance in which explosive/incendiary/chemical elements are placed for the purpose of constituting a device.

Container (Freight) An article of transport equipment which is (1) of a permanent character and strong enough for repeated use; (2) specifically designed to facilitate the carriage of goods by one or more modes of transport without intermediate reloading; (3) fitted with devices permitting its ready handling, particularly its transfer from one mode to another. This definition does not include vehicles. Also: freight container, cargo container, intermodal container.

Container Appurtenances Items connected to container openings needed to make a container a gastight entity. These include safety relief devices, shutoff, backflow check, excess flow check and internal valves; liquid level gauges, pressure gauges and plugs.

Container Assembly An assembly consisting essentially of the container and fittings for all container openings. These include shutoff valves, excess flow valves, liquid level gauging devices, safety relief devices and protective housing.

Container Car A flat or open top RR car such as a gondola, on which containers of freight are loaded.

Container Chassis A trailer chassis having simply a frame with locking devices for securing and transporting a container as a wheeled vehicle.

Container Specification Number Number found on a shipping container preceded by the initials DOT, which indicates the container has been constructed to federal specifications.

Containership A ship specially equipped to transport large freight containers in horizontal, or more commonly, in vertical container cells. The containers are loaded and unloaded by special cranes.

Containment Shell A gas-tight enclosure around a nuclear reactor or other nuclear facility designed to prevent fission products from escaping to the atmosphere.

Contaminant A toxic material found as a residue in or on a substance where it is not wanted.

Contaminate To add an unwanted material, such as a pesticide, which may do harm or damage; to pollute, make impure, or make unfit for use.

Contaminated-Exhaust System An air-cleaning system designed to remove harmful or potentially harmful particulates, mists, or gases from the air exhausted from an operating area.

Contamination The introduction of an unwanted substance into any material or the air. Radioactive contamination is harmful to people and the environment; contaminants affect fire behavior when introduced to normally stable fuels.

Contingency Plan A preplanned document presenting an organized and coordinated plan of action to limit potential pollution in case of fire, explosion or discharge of hazardous materials; defines specific responsibilities and tasks.

Continuous Air Monitor An instrument to detect airborne particulate and gaseous radioactivity and to alarm at specified concentrations.

Continuous Seals A RR term denoting that the seals on a car have remained intact during the movement of the car from point of origin to destination; or, if broken in transit, that it was done by proper authority and without opportunity for loss to occur before new seals were applied. *see* SEALS.

Control Agents Any material that is used to contain or extinguish a hazardous material or its vapors.

Controlled Area Any area controlled by the user for purposes of radiation safety.

Controlled Point A location where the signals and/or switches of a

RR traffic control system are operated and/or controlled from a distant location by a train dispatcher.

Controlled Siding A siding governed by signals under the control of a train dispatcher or operator.

Convection Method of heat transfer whereby the heated molecules circulate through the medium (gas or liquid).

Converter Dolly An auxiliary undercarriage assembly with a fifth wheel and towbar used to convert a semitrailer to a full trailer; also: dolly or converter gear.

Coordination Systematic analysis of a situation and available resources, developing relevant information to inform appropriate command authority of viable alternatives for the most effective actions to meet specific objectives.

Cooperating Agency (ICS) An agency supplying assistance other than direct suppression, rescue, support or service functions to the incident control effort.

Corner Fittings Strong metal devices, located at the corners of a freight container, having several apertures which normally provide the means for handling, stacking and securing the container.

Corner Structures Vertical frame components located at the corners of a container, integral with the corner fittings.

Cornered RR term to denote that a car has been struck by another RR car because it was not in the clear.

Corrective Actions Actions taken by the incident commander to correct the problem at hand in a hazardous materials emergency.

Corrective Action Measures The removal, or treatment in place, of any hazardous constituents that exceed concentration limits in the ground water below a land disposal facility.

Corrosive The quality of a substance which causes the gradual deterioration of another material by chemical processes, such as oxidation or attack by acids.

Corrosive Material Any liquid or solid that can destroy human skin tissue, or a liquid that has severe corrosion rate on steel.

Corrosive Poison A type of poison containing a strong acid or base which will severely burn the skin, mouth, stomach, etc.

Corrosives Substances which attack and "wear away" other materials by strong chemical action.

Corrugated Paper A rigid, structural paper shaped in parallel furrows and ridges.

Cosmic Radiation Radiation originating in outer space.

Coulumb In physics, a unit of electrical charge equivalent to 1 ampere second.

Counter A device measuring ionizations (counts) or particles per unit of time.

Counts Per Minute (CPM) Number of ionizations of particles measured per minute. *see* ACTIVITY, DECAY.

Coupler An appliance for connecting RR cars or locomotives. Government regulations require that these must couple in an automatic manner on impact and must be uncoupled without a person having to move between cars.

Coupler Centering Device (RR) A device for maintaining the coupler normally in the center line of draft but allowing it to move to either side when a car is rounding a curve while coupled to another car.

Coupler Knuckle Lock The block which drops into position when the knuckle closes and holds it in place, preventing uncoupling.

Coupler Lock Lifter The mechanism inside the coupler head moved by the uncoupling rod and in moving, lifts the kunckle lock so the knuckle may open.

Coupling Agent A solvent having the ability to increase the amount of solubility of one material in another.

Coupon Small metal strip which is exposed to corrosive systems in the oil industry for the purpose of determining nature and severity of corrosion.

Covered Gondolas Gondolas equipped with removable cover to protect lading from weather exposure in transit.

Covered Hopper Car A hopper car with permanent roof, roof hatches, and bottom openings for unloading.

CP (RR slang) Abbreviation for CONSTRUCTIVE PLACEMENT or CONTROLLED POINT.

CPCFA California Pollution Control Financing Authority. State agency which helps businesses finance pollution control devices.

Crack a Valve Open a valve slightly so that it leaks just a little; to relieve pressure on a line.

Cradle-to-Grave The tracking of the source, quantity, concentration and type of hazardous waste from generation through final disposal.

Crater To fail, to cave in.

Crew A group of individuals working together as a unit.

Crew Transport Any vehicle capable of transporting personnel in specified numbers.

Crib That portion of ballast between two adjacent RR ties.

Cripple (RR slang) *see* BAD ORDER.

Critical The point at which some quality, property or phenomena undergoes change.

Critical Assembly The assembly of sufficient fissionable material and a moderator to sustain a fission chain reaction at a very low power level permitting study of the behavior of the components for various fissionable materials in different geometrical arrangements.

Critical Mass Minimum amount of fissionable material capable of supporting a nuclear chain reaction.

Critical Point The point at which, in terms of temperature and pressure, a fluid cannot be distinguished as being either a gas or a liquid; the point at which the physical properties of a liquid and a gas are identical.

Critical Pressure That pressure at which a gas becomes a liquid.

Critical Temperature That temperature at which a gas becomes a liquid.

Critical Velocity The lowest water current velocity which will cause loss of oil under the skirt of a containment boom.

Criticality The state of sustaining a chain reaction, as in a nuclear reactor.

Cross Tie Transverse member of RR track structure to which rails are spiked to provide proper gage and to cushion, distribute and transmit the stresses of traffic through the ballast to the roadbed.

Crossing A RR structure used where one track crosses another at grade, consisting of four connected frogs.

Crossmember Transverse members in an underframe.

Crossmember Spacing Distance between crossmember centers.

Crossover (1) Two turnouts with RR track between connecting two nearby and usually parallel tracks. (2) A connection between two channels or passages, such as used in the piping of a flow system.

Crossover Joint A length of pipe casing with one thread on the field end and a different thread in the coupling, used to make a changeover from one thread to another in a string of casing.

Crossover Line A tank piping system to allow unloading from either side of tank.

Crude Oil Petroleum in its natural form before subjected to any refining process.

Crummy Caboose.

Cryogenic Pertaining to liquified gases stored at temperatures approaching absolute zero. Normally, they have a boiling point of about −100°C.

Cryogens Gases that must be cooled to a very low temperature in order to bring about a change from a gas to a liquid.

Cubic Capacity Useful internal load-carrying space, usually expressed in cubic feet, yards, or gallons, carrying capacity of a RR car according to measurement in cubic feet.

Cullet Crushed glass used in glass-making. Use of cullet speeds up melting of silica sand, reducing the energy needed in the manufacturing process.

Cumulative Effect The result of some poisons which build up or are stored in the body so that small amounts contacted over a period of time can sicken or kill a person or animal.

Cupola Small cabin built on roof of a caboose to afford a means of lookout for the train crew.

Curbside Side of a trailer nearest the curb when trailer is traveling in a normal forward direction (right-hand side). Opposite of roadside.

Curie The basic unit to describe the intensity of radioactivity in a sample of material. The curie is equal to 37 billion disintegrations per second, which is approximately the rate of decay of 1 gram of radium.

Current of Traffic Movement of trains on a track in a designated direction specified in the timetable.

Cushion Underframe A term designating the framework of a railway car which is designed to prevent shocks and impact stresses from damaging the car structure or its lading.

Customary A system of measurement used in the United States. Units of the customary system are feet, yards, ounces, pounds, and various units for volume (cups, pints, quarts, gallons).

Cut (1) (Verb). To uncouple a RR car. (2) (Noun). A group of RR cars coupled together. (3) (Noun). That part of the right-of-way excavated out of a hill instead of running up over it or being tunneled through it.

Cut Oil Oil containing water; also called wet oil.

Cutout Cock (Air brake) A valve which will bypass or cut out the brake system for a car when closed. Closing of this valve does not interfere with braking operation on other cars in the train.

Cyclonite A powerful high-explosive material used as the main charge in jet-perforating guns; also called RDX.

Cyclotron A particle accelerator in which charged particles receive synchronized accelerations by electrical fields as the particles spiral outward from their source. The particles are kept in the spiral by a powerful magnetic field.

Cylinder A portable container constructed to DOT cylinder specifications or, in some cases, constructed in accordance with the ASME Code of a similar size and for similar service. Maximum size permitted under DOT specs is 1,000 pounds water capacity.

D

Damage Harm to an inanimate object which disrupts its intended functional continuity.

Damage Free Car RR car equipped with special bracing devices to decrease the possibility of damage to lading; DF car.

Danger Signal word used on a pesticide labels to denote high toxicity (Toxicity Category I) as defined by FIFRA (amended); always accompanied by skull and crossbones symbol and the word "Poison."

Dangerous Cargo Manifest Manifest used on ships containing a list of all hazardous materials on board and their location on the ship.

Daughter Nuclide formed by the radioactive decay of another nuclide, which in this context is called the parent.

Deadhead RR Train and/or engine crew moved without performing service from one terminal to another at railroad convenience and for which they are paid.

Deadman A buried timber, log or beam designed as an anchorage to which a guy wire or cable is fastened to support a structure such as a wood or steel column, derrick, or mast.

Deadman Control (RR) A foot pedal or brake valve which must be kept in a depressed position while the locomotive is operating. A release from this position initiates an air brake application after a short time delay.

Dead Rail A second set of RR tracks over a scale used when cars are not being weighed.

Dead Well An oil well that will not flow.

Debris Any material resulting from the demolition of any structure, including stones, bricks, rocks, concrete, gravel or earth.

Debug To detect, locate and remove mistakes from a routine or malfunctions from a computer.

Decay (Radioactive) The spontaneous transformation of one nuclide into a different nuclide or into a different energy state of the same nuclide. The process results in a decrease, with time, of the number of the original radioactive atoms in a sample. It involves the emission from the nucleus of alpha particles, beta particles (or electrons), or gamma rays; the nuclear capture of ejection of orbital electrons; or fission. Radioactive disintegration.

Decay Product A nuclide, either radioactive or stable, resulting from the disintegration of a radioactive material.

Decay Rate The decrease in activity of a radioactive material within a given time; usually expressed in terms of the period during which half of the atoms will disintegrate (i.e., half-life).

Dechlorination An experimental hazardous waste chemical treatment process which produces a change in the carbon-chlorine bond in organic compounds high in chlorine (e.g., PCB, kepone) with the use of reducing agents.

Decomposition A change in the composition of organic matter to a less complex form; may be accomplished by introduction of heat, through addition of neutralized chemicals, or through biodegradation.

Decomposition Gases Vapors produced by the decomposition of garbage or organic matter. Some are combustible, such as methane.

Decomposition Product Material produced by the physical or chemical degradation of a parent material.

Decompression Very rapid (explosive) reduction of pressure in a pressurized compartment to that of the external atmosphere.

Decontaminate To make safe, purify, make fit for use again by removal of any unsafe substances, such as pesticides or radioactivity.

Decontamination The process of making any person, object, or area safe by absorbing, destroying, neutralizing, or making harmless by removing biological, chemical, or radioactive substances.

Deep-well Injection The disposal of hazardous wastes by pumping into deep wells so they can percolate through porous or permeable subsurface rock and be contained within surrounding layers of impermeable rock or clay. Used extensively for disposal of oil field wastes.

Defect Card (RR) A record keeping device employed by train crews to report problems with the cars in a train. The document is forwarded to the closest repair track for resolution of the problem.

Defect Card Receptacle Small metal container placed underneath the RR car for protection from the weather in which defect cards are placed.

Definitive Care (EMS) Medical Treatment that should remedy an illness or injury.

Deflagration An exothermic reaction in a material which propagates from the burning gases to the unreacted material by conduction, convection and radiation. In this process the reaction zone progresses at a rate less than the velocity of sound.

Deflocculating Agent An adjuvant which prevents precipitation or settling of solids in the suspension fluid.

Defoamer Any chemical that prevents or lessens frothing or foaming in another agent.

Defoliant A pesticide causing the leaves of a plant to drop off.

Defoliate To strip of leaves; to apply a defoliant.

Degradability Ability of a chemical to break down into less complex compounds or elements.

Degradation The breakdown of a more complex chemical into a less complex form; process can be a result of the action of microbes, water, air, sunlight, or other agents.

Degrade To decompose; break down.

De-inking A process in which most of the ink, filler and other extraneous material is removed from printed waste paper, producing pulp which can be used with varying percentages of virgin material in the manufacture of new paper.

Delayed Action In pesticides, referring to certain herbicides producing a delayed response. Considerable time may elapse before maximum effects can be observed, although treated plants usually stop developing soon after treatment and gradually die.

Delisting a Waste The process by which a particular facility proves to DER or EPA that waste from that facility is fundamentally different from waste listed as hazardous and does not meet any of the criteria under which it was listed. The waste product may be excluded from hazardous waste regulation if it does not meet any of the criteria of a hazardous waste.

Delivering Carrier The railroad which delivers a shipment to the consignee.

Delta Designates a difference in something, as in P for difference in pressure.

Deluge Set A device used in fire suppression to apply large volumes of water (master streams) to a fire.

Demurrage A penalty charge assessed by railroads for the detention of cars by shippers or receivers of freight beyond a specified free time.

Demyelination A chemically produced condition which removes or severely damages the myelin sheath around the spinal cord and nerves.

Density (1) The ratio of the weight of a mass to the unit of volume. (2) Number of resident and non-resident persons per square mile.

Density-Dependent Processes in an environment which destroy an increasing percentage of the subject population as the numerical population density increases.

Density-independent Processes in an environment which destroy a relatively constant percentage of the subject population regardless of changes in its density.

Depleted Uranium Uranium having a smaller percentage of uranium 235 than the 0.7% found in natural uranium; obtained from the spent or used fuel elements as by-product tailings or residues of uranium isotope separation.

Depressed Center Flat Car A RR flat car with the section of the floor between the trucks depressed to permit loading of high shipments within overhead clearance limits; a well flat car.

Derail A RR track safety device designed to guide a car off the rails at a selected spot as a means of protection against collisions or other accidents; commonly used on spurs or sidings to prevent cars from fouling the main line.

Dermal Through or by the skin; of or pertaining to the skin.

Dermal Toxicity The degree of a poison absorbed through the skin.

Designated Facility (EMS) A hospital established by local authority to perform specified emergency medical services systems functions.

Dessicant A pesticide used to draw moisture from a plant, plant part, or insect. Used primarily for pre-harvest drying of actively growing plant tissues when seed or other plant parts are developed but only partially mature; or for drying of plants which normally do not shed their leaves, such as rice, corn, small grains, and cereals.

Dessication Dehydration (removal of tissue moisture) by chemical or physical action.

Destination The place to which a RR shipment is consigned.

Detailed Survey The measurement and evaluation of radiation hazards in a given area.

Detection Monitoring Program Procedure utilized to ensure discovery of any leakage from a land treatment facility.

Detonation (1) An exothermic reaction that is characterized by the presence of a shock wave in the material that establishes and maintains the reaction, propagating at a rate equal to or exceeding the speed of sound. (2) A hazardous waste chemical treatment method which treats explosive waste by rapid combustion; explosive destruction.

Detoxify To make harmless; to take out, take away, or neutralize a poison; to remove a poisonous effect.

Detritus Loose material resulting from rock disintegration or abrasion; suspended material in the water column including fragments of decomposing flora and fauna, and fecal pellets produced by zooplankton and associated bacterial communities.

Deuterium An isotope of hydrogen whose nucleus contains one neutron and one proton, making it about twice as heavy as the nucleus of normal hydrogen, which is only a single proton. Often called heavy hydrogen, it occurs in nature as 1 atom to 6,500 atoms of normal hydrogen and is not radioactive.

Dialysis Process of separating a mixture of substances in solution by using a membrane as a filtering agent. Substances move through the membrane at varying rates and separate according to their relative molecular weights.

Diamond RR crossing.

Diesel Electric Locomotive A train engine in which one or more diesel engines drive electric generators which in turn supply electric motors (usually series D.C.) which are geared to the driving axles.

Diffusion (1) Mixing of substances, usually gases and liquids, due to molecular motion. (2) The spreading out of a substance to fill a space.

Diffusion Plant (Gaseous) A method of isotopic separation based on the fact that gas atoms or molecules with different masses will diffuse through a porous barrier (or membrane) at different rates. Method is used by the AEC to separate uranium 235 from uranium 238 and requires large gaseous-diffusion plants and enormous amounts of electric power.

Dike An embankment of natural or man-made materials constructed to contain or obstruct the movement of liquids, sludges, or other substances.

Diluent Any liquid or solid material used to weaken or to carry an active ingredient such as a concentrated pesticide.

Dilute To weaken a solution by adding water, oil, or other liquid or solid.

Dinky A small RR engine used around roundhouses or backshops for switching.

Dip The complete or partial immersion of a plant, animal or object in a pesticide.

Dip Tube Device for pressure unloading of a product from the top of a tank.

Direct-Cycle Reactor System A nuclear power plant system in which the coolant or heat transfer fluid circulates first through the reactor and then directly to a turbine.

Direct Gas-Fired Tank Heater A device which applies heat from a gas burner flame directly to a portion of a container surface in contact with LP-Gas liquid to vaporize the product at the rate needed to supply the connected gas-consuming devices.

Directed Application Aiming a pesticide at a specific area such as at a portion of a plant, animal, or structure or at a row or bed.

Disaster A perilous condition which exceeds a jurisdiction's capability to control with immediately available resources.

Disaster Medical Services (EMS) Those medical services provided to disaster victims which minimize morbidity and mortality.

Disaster Operations Center (DOC) Location for the coordination of disaster response activities by a state Department of Health Services.

Disaster Response Component (EMS) Those resources and arrangements necessary to adequately respond to mass casualty situations on a local, regional or statewide basis.

Disaster Support Area (DSA) (EMS) A designated area on the periphery of an emergency incident from which relief resources can be received, stored, allocated and dispatched.

Discharge An intentional or accidental spilling, leaking, pumping, pouring, dumping, emitting or any other release of hazardous waste, hazardous waste constituents, or hazardous materials which, when released into land or water, become hazardous waste.

Discretionary Project An undertaking requiring the exercise of judgment when a public agency decides to approve or disapprove a particular activity; as distinguished from situations where that agency merely has to determine if there has been conformity with applicable statutes, ordinances or regulations.

Discriminator An electronic circuit which selects signal pulses according to their pulse height or voltage. Used to delete extraneous radiation counts or background radiation or as the basis for energy spectrum analysis.

Disinfectant Pesticide that controls germs.

Disinfection Effective killing by chemical or physical processes all organisms capable of causing infectious disease. Example: chlorination is commonly employed in sewage treatment processes.

Disintegration A nuclear transformation; to break or decompose into constituent elements.

Dispatch (ICS) Implementation of a decision to move resources from one place to another.

Dispensing Device (Dispenser) A device normally used to transfer and measure LP-Gas for engine fuel into a fuel container, serving the same purpose for an LP-Gas service station as that served by a gasoline dispenser in a gasoline service station.

Dispersants Chemicals which reduce the surface tension between oil and water, thereby facilitating the breakup and dispersal of an oil slick throughout the water column in the form of an oil-in-water emulsion. Chemical dispersants can only be used in areas where adverse biological damage will not occur and then only when approved for use by government regulatory agencies.

Dispersing Agent An adjuvant that reduces the attraction between particles.

Dispersion The distribution of spilled oil into the upper layers of the water column by natural wave action or application of chemical dispersants. In shoreline cleanup and restoration, the removal of stranded oil through wave action (natural dispersion), or application of chemical dispersants, or use of one of various hydraulic dispersion techniques.

Dispersion Mechanism The process by which a material scatters in one or more directions.

Disposable Designed to be used for a brief period of time, then discarded; neither durable nor repairable.

Disposal Well An oil well through which water (usually salt water) is returned to subsurface formations.

Dissolution The process of dissolving one substance in another; in oil spills, a process contributing to the weathering of spilled oil whereby certain slightly soluble hydrocarbons and various mineral salts present in oil are dissolved in the surrounding water.

Distributing Plant (LP-Gas) A facility which receives LP-Gas in tank car, truck transport or truck lots, distributing this gas to the end user by portable container delivery, tank truck or through gas piping. Such plants have bulk storage (2,000 gallons water capacity or more) and usually have container filling and truck loading facilities on the premises; bulk plant. Normally no persons other than the plant management or plant employees have access to these facilities.

Distributing Point (LP-Gas) A facility other than a distributing plant normally receiving LP-gas by tank truck, which fills small containers or the engine fuel tanks of motor vehicles on the premises. Any such facility having LP-Gas storage of 100 gallons or more water capacity and to which persons other than the owner of the facility or his employees have access, is considered to be a distributing point; an LP-Gas service station.

Division (RR) That portion of a railroad assigned to the supervision of a superintendent, usually consisting of yards, stations and sidings.

DOE United States Department of Energy.

DOHS State Department of Health Services; state agency whose responsibilities include toxic waste management and regulation in California.

DOLLY *see* CONVERTER DOLLY.

Dome The circular fixture on top of a tank car containing valves and relief devices.

Dormant Spray A pesticide application made before trees and other plant life begin to leaf out in the spring.

Dose (1) The amount of ionizing radiation energy absorbed per unit mass of irradiated material at a specific location, such as a part of the human body, measured in REMS, or an inanimate body, measured in rads. (2) Amount, quantity, or portion of a pesticide which is applied to a target. Also, a consistent measure used in testing to determine acute and chronic toxicities.

Dose Equivalent Amount of effective radiation when modifying factors have been considered. The product of absorbed dose multiplied by a quality factor multiplied by a distribution factor; expressed numerically in REMS.

Dose Projections A calculated estimate of the potential dose to individuals at a given location; projection is determined from the quantity of radioactive material released from the source and the appropriate meteorological transport and dispersion parameters.

Dose Rate The radiation dose delivered per unit time and measured in rems per hour.

Dosimeter Radiation measuring device, such as a lapel badge or an ionization chamber.

Dosimeter Charger Device utilized to induce a static electric charge on a dosimeter.

DOT United States Department of Transportation.

Double (1) Two consecutive tours of duty. (2) Putting train together when part of train is on one track and balance on another.

Doubles Trailer combination consisting of a truck tractor, semi-trailer, and a full trailer coupled together.

Double Containment Arrangement of double barriers whereby the second barrier provides backup protection against leakage through, or failure of, the first.

Double Deck (Stock car) A RR car with a second floor (often removable) halfway between the ordinary floor and roof to increase carrying capacity of car for small livestock.

Double Filtration (1) Arrangement of double barriers in which the second provides protection against leakage of the first. (2) A series arrangement intended to increase total filtration efficiency.

Doughnut A ring of wedges supporting a string of pipe, or a threaded, tapered ring used for same purpose.

Downwind Direction toward which the prevailing wind is blowing.

Draft Gear Unit on a RR car which forms connection between the coupler rigging and the center sill and cushions force of impact of shocks incidental to train movements.

Drag (1) A train of empty RR cars. (2) A heavily-laden train. (3) Group of cars for movement from one point to another within a terminal.

Drawbar Coupler.

Drawhead Head of an automatic coupler, exclusive of the knuckle, knuckle pin and lock.

Dresser Sleeve A slip-type collar used to join plain-end pipe.

Drift Movement of particles by wind and air currents from the target area to an area not intended to be treated.

Drill Track Track connecting with the ladder track, over which engines and cars move in switching.

Drilling (Car) Handling or switching of RR cars in freight yards.

Drinking Water Supply Any raw or finished water source that is or may be used as a public water system or as drinking water by one or more individuals.

Drome Box On tractor-trailer vehicles, a large boxlike structure between the cab and the trailer, often containing hazardous materials.

Drop RR switching movement when cars are cut off from an engine and allowed to roll free into a track.

Drop-Axle An axle in which the centerline of the beam portion is offset from the centerline of the spindles.

Drop Bottom Car A gondola car with a level floor, equipped with drop doors for discharging the load.

Drop Brake Shaft A brake shaft for flat cars which normally extends above the floor, but can be dropped down should conditions of the lading require.

Drop-End Gondola Car Gondola car with end doors which can be dropped for shipping long material extending over more than one car.

Drop Frame A two-level section of trailer providing proper coupler height on the forward end and a lower floor height for the remainder of the trailer length.

Droplet Breakaway A type of boom failure resulting from excessive current velocity; the head wave formed upstream of the mass contained within a boom becomes unstable.

Dry Gas Natural gas produced without liquids; also, a gas that has been treated to remove all liquids.

Dummy Coupler A dust cap fitting used to seal the opening in an air brake hose connection when not in use.

Dump An open land site where waste materials are disposed of in a manner which does not protect the environment, is susceptible to open burning, or is exposed to the elements, vermin and/or scavengers.

Dump Car A RR car capable of discharging load either through doors or by tipping the car body.

Dunnage The material used to protect or support freight in or on cars, such as bracings, false floors, meat racks, etc.

Durable Long-lasting in spite of hard wear or frequent use.

Dust A finely ground dry mixture combining a small amount of pesticide with an inert carrier such as clay, talc, or volcanic ash.

Dutchman A piece of pipe that has been twisted off inside a female connection; a short section of material, such as belting or pipe, used to lengthen existing equipment.

Dynamic Braking Using the motors of a locomotive as generators and dissipating power through resistors to control train speed, after which air brakes are utilized to bring the train to a full stop.

Dzus Fastener Trade name for a screw-like fastener with a slotted head used on aircraft. One-half turn either way secures or releases the fastener.

E

Easy Sign Hand signal indicating the train or other vehicle is to move slowly.

EC50 The median effective concentration (ppm or ppb) of the toxicant in the environment (usually water) producing a designated effect in 50 percent of the test organisms exposed.

Ecology Study of the relationship between a plant or animal and its surroundings.

Economic Poison Term formerly used for pesticide.

Ecosystem Interacting system of a biological community and its non-living environment.

ED50 The median effective dose (usually expressed as mg/Kg or mg/g of body weight) producing a designated effect in 50 percent of the exposed test organisms.

Effective Concentration (EC) The concentration of a chemical substance effective in producing a specific result such as increase in oxygen consumption, paralysis, death, etc.

Effluent (1) Discharge or outflow of water from ground or subsurface storage. (2) Solid, liquid, or gas wastes which enter the environment as a by-product of man-oriented processes.

Electric locomotive A locomotive which receives power from an overhead contact wire or third rail and uses that power to drive electric motors connected by gears to the driving axles.

Electrically Locked Switch Hand-operated switch equipped with an electrically controlled device which restricts the movement of the switch.

Electrically Neutral Term given to an object with a net electrical charge of 0. Example, an atom of hydrogen has 1 proton (+1) and one electron (−1); the two opposite charges cancel, and the atom is neutral.

Electrodialysis Process of separating substances in a solution by dialysis using an electric field as the driving force.

Electrolysis A hazardous waste chemical treatment method by which chemical changes are accomplished at the surface of electrodes carrying an electric current and immersed in a chemical solution.

Electrolyte (1) A group of chemicals essential to biological functions. In solution, the molecules dissociate into ions and aid in transferring electrical impulses in the nervous system. (2) A substance that, when dissolved in a suitable solvent or when fused, becomes an ionic conductor.

Electromagnetic Radiation Radiation consisting of associated and interacting electric and magnetic waves that travel at the speed of light.

Electron A component of an atom; travels in a distant orbit around a nucleus.

Electrostatic Precipitator Device on smokestacks collecting small particles of dust by giving them an electrical charge and attracting them to a collecting electrode.

Element (1) The most simple substance that cannot be separated into more simple parts by ordinary means. There are approximately 100 elements. (2) A group of atoms which all display the same physical and chemical properties and have the same number of protons in their nuclei. (3) (ICS) Any identified part of the Incident Command System organizational structure.

Elementary Particles The simplest particles of matter and radiation.

Elevated Release A release of radioactive effluents via the ventilation release point on a reactor containment building, including effluents from containment building and a supplementary leak collection and release system.

Embargo Order prohibiting acceptance and/or handling of freight at certain points or via certain routes due to emergencies, congestion, strikes, etc.

Emergency A situation of concern which has developed suddenly and unexpectedly, and which demands prompt action.

Emergency Actions A collective term encompassing the assessment, corrective, and protective actions taken during the course of an emergency.

Emergency Application A quick, heavy reduction of brake pipe pressure made when a train must be stopped in the minimum distance possible.

Emergency Brake Valve A valve for applying the train brakes in emergency, connected to the brake pipe by a branch pipe and operated by releasing brake pipe air to the atmosphere.

Emergency Communications Center (EMS) A facility designated by a political jurisdiction as responsible for receiving and transmitting emergency communications.

Emergency Coordinator The individual in an organization who has been appointed in a preplan as responsible for direction of on-site actions during an emergency.

Emergency Dose Limit A level of projected absorbed dose following a nuclear incident above which the total risk to health of an individual is considered excessive.

Emergency Implementing Procedures Preplanned detailed actions to be taken in case of emergency incident.

Emergency Kit A container of equipment and supplies needed during a serious emergency and reserved exclusively for that purpose.

Emergency Measures Collective term encompassing the assessment, corrective, and protective actions taken during the course of the emergency condition.

Emergency Medical Services (EMS) Functions required to provide urgent medical care for ill or injured patients, such as communications, transportation, medical personnel and/or administration.

EMSA Emergency Medical Services Authority.

EMS System A plan providing personnel and resources to respond to the emergency medical needs of ill and/or injured persons within a defined geographic area.

Emergency Medical Technician (EMT) An individual trained in all facets of basic life support according to standards prescribed by a State and who has a valid certificate issued pursuant to state law.

Emergency Operations Center A location at the headquarters of each off-site response agency from which control and/or coordination of off-site emergency actions are directed.

Emergency Plan A preplanned outline of what protective measures should occur in the event of any sort of hazardous incident.

Emergency Radiation Monitoring Systems (ERMS) An instrument system at a nuclear facility designed to detect and alarm if high level radiation emanates from the containment building.

Emergency Rate (Freight) A RR rate established to meet some immediate and pressing need, without due regard to the usual rate factors.

Emergency Reservoir A part of the AB air brake system to provide quick recharge, graduated release and high emergency cylinder pressure, supplementing the supply from the auxiliary reservior.

Emergency Response Action taken to ameliorate the consequences of an emergency situation.

Emergency Shut-Off Levers A means of operating a valve, usually manually that stops the flow of a liquid when the normal shut-off is not available because of a leak or spill.

Emergency Siren A signal generator sometimes coupled to a horn, operated either manually or automatically, used to alert personnel to a prearranged response pattern.

Emergency Valve A self-closing tank outlet valve.

Emergency Valve Remote Control A secondary closing means, away from tank discharge openings, for operation in event of fire or other accident.

Eminent Domain The right of a government to appropriate private property for necessary public use with compensation paid to the landowner.

Empty Car Bill Waybill used to move ordinary empty RR cars from one station to another.

Emulsifiable Concentrate Solution A solution containing high concentration of active ingredient to be mixed with water.

Emulsification Process whereby one liquid is dispersed into another liquid in form of small drops.

Emulsifier (Emulsifying agent) A chemical which helps one liquid form tiny droplets and thus remain suspended in another liquid.

Emulsion A mixture in which one liquid is suspended as tiny droplets in another liquid.

Encapsulation (1) Method of formulating pesticides in which the active ingredient is encased in material (often plastic) resulting in sustained pesticidal release and decreased hazard. (2) Method of disposal of pesticides and pesticide containers by sealing them in sturdy, waterproof, chemical-proof containers which are then sealed in thick plastic, steel, or concrete to resist damage or breakage so the contents cannot get out.

End Door (Lumber door) A door in the end of a RR car.

Endangered Persons People in the exposure area created by a hazardous materials incident.

Endolytic Insecticides Systemic insecticides remaining in their original form until decomposed by the biological system.

Endometatoxic Insecticides Systemic insecticides metabolized as a secondary toxicant by the biological system on which they have been used.

Endothermic A chemical reaction which absorbs heat in the process. Endothermic materials produce materials with more total energy than the reacting substance.

Energy A system capable of producing a physical change of state.

Energy Recovery Obtaining energy from the controlled incineration of solid waste.

Energy Sources Chemical, electrical, mechanical, thermonuclear.

Energy Transfer Changing a state of being.

Engine A unit propelled by any form of energy or a combination of such units operated from a single control.

Enriched Material Substance in which the percentage of a given isotope present has been artificially increased so that it is higher than the percentage of that isotope naturally found in the material.

Entomologist Scientist specializing in the study of insects.

Environment (1) The air, water and the earth, sometimes called the biosphere. (2) The sum total of all the external conditions that may act upon an organism or community to influence its development or existence.

Environmental Impact Report (EIR) A detailed statement describing and analyzing the significant environmental effects of a project and discussing ways to mitigate those effects.

Environmental Protection Agency (EPA) A federal agency charged with implementing the Resource Conservation and Recovery Act and having general responsibility to administer programs which address the environmental problems of water and air pollution, toxic substances, pesticides, radiation, noise, and solid waste management.

EPA Establishment Number An identification number assigned to each pesticide production plant by the EPA which must appear on all labels of that product.

EPA Registration Number Number appearing on a pesticide label to identify the individual pesticide product; may appear as "EPA REG NO."

Environmental Sensitivity The susceptibility of a local environment or area to any disturbance which might decrease its stability or result in either short- or long-term adverse impacts. Environmental sensitivity generally includes physical, biological and socioeconomic parameters.

Environmentally Sensitive Area. Areas exceptionally responsive to environmental change and especially prone to irreversible ecological upset. These can include wetlands, floodplains, permafrost areas, critical habitats of endangered species, and recharge areas of aquifers.

EP Toxicity A characteristic indicating the likelihood that certain constituents could be leached by a medium in significant concentrations, as determined by a specific extraction procedure.

Equalizer Beam Suspension device used to transfer and maintain equal load distribution between two or more axles of suspension; rocker beam.

Equalizer Hanger Bracket for mounting equalizer beam of multiple axle suspension to trailer frame, which allows for beam's pivotal movement; center hanger.

Eradicant A pesticide which kills the pest after it appears.

Eradicant Fungicide Pesticide which kills a fungus after it appears on or in a plant.

Erythema Abnormal redness of the skin, as in inflammation.

Erythrocytes Red blood cells.

Estuary Widened channel of the mouth of a river in which influences of the tides is felt; delicate ecosystems serving as nurseries, spawning and feeding grounds for large groups of marine life and provide shelter and food for birds and wildlife.

Etiologic Agent A living microorganism that may cause human disease; germs.

Evacuation Removal of residents from an area of danger.

Evacuation Area Area surrounding an emergency incident in which personnel and/or residents may be subject to ready removal in case of necessity.

Evaporate To form a gas and disappear into the air; to vaporize.

Evaporation The process of a solid or liquid turning into a gas. Also, a hazardous waste physical treatment process by which suspended and dissolved solids are separated from liquid waste by evaporation of the liquid.

Event An occurrence, act or decision or other occurrence in an ordered sequence of occurrences; connotes action, dynamic rather than static.

Events Sequence A series of events in a logical relationship.

Exceptions to Classification A RR publication containing classification ratings (a percentage of first class) and rules different (generally lower) from those shown in the Uniform Freight Classification.

Excess Flow Valve Device designed to close when the liquid or vapor passing through exceeds a prescribed flow rate as determined by pressure drop.

Exclusion Area Area surrounding a particular incident site, specifically a nuclear reactor, wherein all unauthorized personnel may be prohibited.

Excursion A sudden, very rapid rise in the power level of a reactor caused by supercriticality. Excursions are usually quickly suppressed by the negative temperature coefficient of the reactor or by automatic control rods, or both.

Exemption An exception to a policy, rule, regulation, law or standard.

Exercise A realistically planned simulation of an accident, designed and coordinated in such a manner that the response of the emergency organization and other station personnel closely approximates the response to an actual incident. An exercise may involve participation of off-site organization(s).

Existing Hazardous Waste Management Facility Any storage, treatment or permitted disposal facility which was in operation on November 29, 1980, or for which construction was begun on or before November 19, 1980.

Existing Portion That land surface area of an existing waste management unit, upon which wastes have been placed prior to the issuance of a permit.

Exoskeleton The segmented external skeleton of an insect; the insect's "skin."

Exothermic A chemical reaction which gives off heat. Exothermic materials produce matter with less total energy than the reacting substances.

Expansion Loop A bend placed in a line to absorb stretch or shrinkage.

Explosion Sudden release of large amount of energy in a destructive manner; a result of powders, mists or gasses undergoing instantaneous ignition, or liquids or solids undergoing sudden decomposition, or a pressurized vessel undergoing overpressure rupture with such force as to generate tremendous heat, cause severe structural damage, occasionally generating a shock wave, and propelling shrapnel.

Explosive A term applied to solid or liquid substances possessing the faculty under certain circumstances of undergoing instantaneous decomposition, extending throughout their entire mass. The process is accompanied by a considerable disengagement of heat, the substance being wholly or partially converted into gaseous products.

Explosive Range (Flammable range) The boundary lines of mixture or concentration of flammable vapor and air, which, if ignited by an outside heat source, will propagate flame.

Export To send goods to a foreign country.

Exposure Contact of living tissue with ionizing radiation. *see CONTAMINATION.*

Exposure Rate The rate of exposure to ionizing radiation, measured in roentgens/hour.

Extended Planning Zone An area beyond the emergency zone and out to approximately twenty miles in which extensive planning (but not evacuation) for public protection is accomplished.

Extra Train Train not included in a timetable schedule.

Extremely Flammable A liquid that has a flash point of 20°F or lower, determined by closed cup of Seta flash test.

F

F Fahrenheit degrees.

Fabrication in Transit The stopping of steel products at a point located between the points of origin and destination for further process or manufacture. For example, steel beams to be fabricated as bridge girders.

Facility The on-site structures and all adjoining land and rights of way which are used for testing, storing, or disposing of hazardous wastes.

Facing Movement The movement of a train over the points of a switch which face in a direction opposite to that in which the train is moving.

Fecal Pellets Solid or semi-solid excretion products (faeces) which are enclosed within a thin membrane, such as produced by zooplankton and some other invertebrates.

Failure Mode The way in which a failure occurred.

Fallout Airborne particles containing radioactive material that fall to the ground following a nuclear explosion.

Fatigue Failure of a metal under repeated loading.

Fauna Animals in general or animal life as distinguished from plant life (flora); usually used in reference to all the animal life characteristic of or inhabiting a particular region or locality.

FDA Food and Drug Administration of the United States.

Feasibility Study Detailed examination of the technical, environmental, engineering, economic, legal and practical suitability of a proposed facility or technology for use at a specific location.

FEPCA Federal Environmental Pesticide Control Act of 1972, regulated by EPA.

Feedback An element of a system that is the return of a portion of the outputs to the input, thus allowing the system to evaluate itself.

Feeding in Transit The stopping of shipments of livestock to be fed and watered at a point located between the points of origin and destination.

FEMA United States Federal Emergency Management Agency.

Ferrous Metals Metals which are predominantly composed of iron; most are magnetic.

FHWA Federal Highway Administration; DOT division concerned with highway construction and usage. Other divisions of DOT relate to air, rail and water transportation.

Field An area consisting of a single oil reservoir or multiple reservoirs all grouped on, or related to, the same individual geological structural feature and/or stratigraphic condition.

Field Administrative Limit A pre-established limit on radiation absorbed dose for emergency personnel. The limit is set by local jurisdictions for use in managing field actions in response to radiological incidents.

FIFRA (amended) Federal Insecticide, Fungicide, and Rodenticide Act of 1947 (amended); regulated by EPA.

Fifth Wheel A device used to connect a truck tractor or converter dolly to a semitrailer in order to permit articulation between the units. It generally is composed of a lower part consisting of a trunnion, plate and latching mechanism mounted on the truck tractor (or dolly) and a kingpin assembly mounted on the semitrailer.

Fifth Wheel Pickup Ramp A steel plate designed to lift the front end of a semitrailer to facilitate engagement of kingpin into fifth wheel.

Fill Opening Opening in top of tank used to load container, usually incorporated in manhole cover.

Fill, Filling Transferring liquid LP-Gas into a container.

Filling by Weight *see* WEIGHT FILLING.

Film Badge A light tight package of photographic film worn like a badge by workers in nuclear industry or research and used to measure possible exposure to ionizing radiation. The absorbed dose can be calculated by the degree of film darkening caused by the irradiation.

Filter A device or substance for straining out solid particles or impurities from a liquid or gas.

Filter Bank A parallel arrangement of filters on a common mounting frame enclosed within a single housing.

Filtration A hazardous waste physical treatment process which suspends particles from liquid by forcing fluid through a porous substance such as paper, cloth, fine clay, sand or charcoal, entrapping suspended particles on or within the filter medium.

Final Cover The cover material applied upon closure of a landfill; it is permanently exposed at the surface.

Finite Tolerance The maximum amount of pesticide which can legally remain on a food or feed crop at harvest after the pesticide has been directly applied to the crop.

Fire Rapid oxidation of a fuel. *see* COMBUSTION.

Fire Point The lowest temperature of a liquid at which vapors are evolved fast enough to support continuous combustion. Related to FLASH POINT.

Fireproof A misnomer applied to building materials or structural components of a building.

Firescope (ICS) Firefighting Resources of Southern California Organized for Potential Emergencies.

Fire Wall A wall of earth built around an oil tank to hold the oil if the tank breaks or burns; a berm.

First Responder(s) (EMS) Initial unit dispatched to the scene of a medical emergency to provide patient care.

Fissile Material which is capable of undergoing fission.

Fissile Material Any material fissionable by neutrons of all energies including and especially thermal (slow) neutrons as well as fast neutrons (uranium 235 and plutonium 239); term is a more restricted meaning than fissionable.

Fission A process in which large radionuclides break into smaller pieces and release radiation in the form of particles or energy. The splitting of an atomic nucleus into two parts accompanied by the release of a large amount of radioactivity and heat.

Fission Products The nuclei (fission fragments) formed by the fission of heavy elements, plus the nuclides formed by the fission fragments' radioactive decay.

Fissionable A nucleus that undergoes fission under the influence of neutrons, even of very slow neutrons.

Fissionable Materials Commonly used as a synonym for fissile material. Meaning of this term also has been extended to include material that can be fissioned by fast neutrons only, such as uranium 238. Used in reactor operations to mean fuel.

Fittings The small pipes and valves used to make up a system of piping.

Fixed Liquid Level Gauge A gauge using a relatively small positive shutoff valve and designed to indicate when the liquid level in a container being filled reaches the point of communication with the interior of the container.

Fixed Maximum Liquid Level Gauge A fixed liquid level gauge which indicates the liquid level at which the container is filled to its maximum permitted filling density.

Fixed Signal (RR) A signal of fixed location indicating a condition affecting movement of train or engine.

Flag Station A station at which trains stop only when signalled.

Flame Impingement The points where flames contact the surface of a container.

Flame-Spread Speed at which a flame will cross the surface of a material, influenced by the physical form of the fuel, air supply, fuel's moisture content, specific gravity, size and form, the rate and period of heating, and the nature of the heat source. A higher flame-spread critically affects the severity of the fire in a given period of time.

Flammability, Flammable The capacity of ignition of a substance. Generally, the more flammable a substance the more likely the spread of fire. In building materials, used in a general sense; in Class B liquids, applies to substances with flash points below 140°F. Commodity which can be easily ignited.

Flammable Gas Any gas that will burn.

Flammable Liquid Any liquid with flash point below 100°F (37.7°C).

Flammable Material A substance capable of being easily ignited and of burning rapidly.

Flammable Range Range of concentrations of a flammable vapor in the air. Lower flammable limit marks point where vapors are too lean to burn; above the upper flammable limit the vapors are too rich to burn.

Flammable Solid Any material, other than an explosive, liable to cause fires through friction, retained heat from manufacturing or processing, or that can be ignited readily and when ignited burns so vigorously and persistently as to create a serious transportation hazard.

Flange A rib or rim for strength or guidance. A projecting edge on the circumference of a RR wheel to keep it on the rail.

Flash Point Temperature when a liquid gives off flammable vapors sufficient to form an ignitable mixture near the surface of the liquid; combustion is not continuous at the flash point.

Flashing Liquid-tight rail on top of a tank which contains water and spillage and directs it to suitable drains.

Flashing Drain Tubing which drains water and spillage from flashing to the ground.

Flat Car Open car without sides, ends or top, used primarily for hauling lumber, stone, heavy machinery.

Flexible Connector A short (not exceeding 36″ overall length) component of a piping system fabricated of flexible material such as hose and equipped with suitable connections on both ends; used where there is a possibility of greater relative movement between points connected than is acceptable for rigid pipe.

Float Bridge A bridge connecting car floats with rail landings.

Float Gauge A gauge constructed with a float inside the container resting on the liquid surface which transmits its position through suitable leverage to a pointer and dial outside the container indicating the liquid level.

Floatage The transfer of RR cars across water.

Floating Load (RR) A load in which the lading is prepared as a unit with space between unit and ends of car and end blocking omitted. Lengthwise movement of the lading over floor of the car permits the dissipating of impact shocks.

Flocculation A hazardous waste physical treatment method by which suspended particles are assembled into larger more settleable particles after the waste is mixed with chemicals. This technique enhances the sedimentation process.

Flood Plain Lowland bordering a river which is usually dry but subject to flooding when the stream overflows.

Flora Plants in general; plant life as distinguished from animal life (fauna). Usually used in reference to all the plant life inhabiting or characteristic of a particular region or locality.

Flotation A hazardous waste physical treatment process by which particles are separated from liquid by introducing fine gas bubbles which attach to the particles and rise to the surface; particles are then collected by skimming mechanisms.

Flowable (1) Very finely ground solid material which is suspended in a liquid; usually contains a high concentration of the active ingredient and must be mixed with water when applied. (2) A specialized type of pesticide formulation wherein a very finely ground solid particle is mixed in a liquid carrier, also called a flowable liquid or water-dispersible suspension, consisting of a wettable powder suspended in an oil or liquid base.

Fly Ash Fine particles of ash of a solid fuel which are either carried out of the flue with the waste gases produced during combustion or recovered from the waste gases.

Flycrew (ICS) A handcrew of predetermined size transported to an incident via helicopter.

Foaming Agent A material which causes a pesticide mixture to form a thick foam; used to reduce drift.

Fogger An aerosol generator; pesticide equipment that breaks some pesticides into very fine droplets and blows or drifts the "fog" onto the target areas.

Foliar Application Spraying a pesticide onto the stems, leaves, needles, and blades of grasses, plants, shrubs, or trees.

Food Chain The dependence of one type of life on another, each in turn eating or absorbing the next organism in the chain.

Food-Chain Crops Those crops grown for human consumption and pasture and other crops grown for feed for animals, whose products or by-products will be used for human consumption.

Force A vector quantity of energy that tends to produce an acceleration in the direction of its application.

Fork Pockets Transverse structural apertures in the base of the container which permit entry of fork lift devices.

Formulation The pesticide product; it may contain one or more active ingredients, the carrier (if needed), and other additives (if needed) to make it ready for sale in a form that is safe and easy to store, dilute, and apply.

Formulation Plant Plant designed to combine the technical pesticide with a solvent or diluent in order to prepare the material for commercial use.

Fractional Detonation Partial detonation of a high explosive, resulting in the scattering of device parts and undetonated explosives over a wide area.

Fractional Distillation Separation of a mixture of liquids having different boiling points, a primary process in the refining of crude oils.

Framing Lumber Material used in construction and framing of houses. Usually pressure-treated with pentachlorophenol, a wood preservative.

Freeboard Vertical distance between top of a tank sidewall or lowest elevation of a surface impoundment dike or berm, and elevation of the highest surface of the waste contained in the tank or impoundment. Also, part of a floating boom designed to prevent waves from washing oil over the top.

Free Liquids Liquids which readily separate from the solid portion of a waste under ambient temperature and pressure.

Free Time The time allowed by the carriers for the loading and unloading of freight after which demurrage or detention charges will accrue.

Freight Agent RR representative who prices services performed based on approved tariffs.

Freight Bill Statement given to customer for charges of transportation; information is taken from waybill.

Freight Charge Assessment for transporting freight.

Freight Claim Demand upon a carrier for payment of overcharge, loss, or damage sustained by shipper or consignee.

Freight Classiciation *see* CLASSIFICATION, UNIFORM FREIGHT CLASSIFICATION.

Freight Forwarder An individual or organization engaged in business of shipping and distributing less than a carload of freight.

Freight House The station facility of a transportation line for receiving and delivering freight.

Frog (1) A track structure used at the intersection of two running rails to provide support for wheels and passageways for their flanges, thus permitting wheels on either rail to cross to the other. (2) An implement for rerailing car wheels.

Front-End Loader A refuse truck which has power-driven loading equipment at front of the vehicle.

Fuel Any combustible substance which is burned to produce useful heat energy.

Fuel Element A rod, tube, plate, or other mechanical shape or form into which nuclear fuel is fabricated for use in a reactor.

Fuel Oils Refined petroleum products having specific gravities in the range from 0.85–0.98 and flash points greater than 55°C; includes furnace, auto diesel, and stove fuels, plant or industrial heating fuels and various bunker fuels.

Fuel Tender (ICS) Vehicle capable of supplying fuel to ground or airborne equipment.

Fuel Value Applying to the amount of potential energy to be released by a fuel in the combustion process; expressed in terms of

BTUs per pound of fuel. Examples:

Substance	BTU/pound
Wood Shavings	8,248
Wool Rags	8,876
Cotton Rags	7,165
Asphalt	17,158
LPG	18,000

Full Protective Clothing Clothing that will prevent gases, vapors, liquids, and solids from contacting the skin; includes helmet, self-contained breathing apparatus, coat and pants customarily worn by firefighters, rubber boots, gloves, bands around legs, arms and waist, and face mask, as well as covering for neck, ears, and other parts of head not protected by the helmet, breathing apparatus or face mask.

Full Service Application Brake application resulting from a reduction in brake pipe pressure at a service rate until maximum brake cylinder pressure is developed.

Full Trailer A truck trailer constructed so that all its own weight and that of its load rests upon its own wheels. A semitrailer equipped with a dolly is considered a full trailer.

Fulminate In medicine, to develop suddenly and severely, as in a disease.

Fume A smoke, vapor, or gas.

Fumigant A lethal pesticide in the form of a gas may be a liquid which becomes a gas when applied.

Fumigation Use of chemicals in form of gas to destroy noxious insects, nematodes, and unwanted plants.

Fungi Groups of small plant organisms which cause rots, molds, and plant diseases. They lack cholorophyll, grow from seed-like spores and produce tiny thread-like growths. Some fungi can attack and destroy non-living things. Examples: molds, mushrooms and yeasts.

Fungicide Pesticide that controls or inhibits fungus growth.

Fusee Red flare used for flagging purposes.

Fusible Plugs A safety relief device in form of a plug of a low-melting metal. Plugs close the safety relief device channel under normal conditions and are intended to yield at a set temperature to permit the escape of gas.

Fusion (1) Transition of a material from solid to a liquid (melting). (2) A process in which small nuclei combine into larger nuclei and release radiation in the form of particles or energy.

G

Gallon Unit of Measure, U.S. Standard; 0.833 Imperial Gallons = 231 cubic inches—3.785 liters.

Gamma A type of electromagnetic radiation; a form of ionizing radiation.

Gamma Rays High energy, short wave-length electromagnetic forms, comprised of photons or fine packets of energy, which travel in straight paths at the speed of light; very penetrating but do not make material radioactive. Best shielded by dense materials such as lead or depleted uranium.

Gamma Ray Irradiation Experimental hazardous waste chemical treatment method which disinfects waste by utilizing gamma radiation to destroy pathogens (disease-causing microorganisms).

Gandy Dancer (slang) RR track laborer.

Gantry Crane A stilted traveling crane supported on a bridge or trestle. Trestle belts are constructed on wheels so the whole structure travels on a track laid on the ground or floor.

Garbage Decayable animal and vegetable wastes resulting from handling, preparation, cooking and consumption of food. Also, trash or solid waste; things discarded regardless of reusability or recyclability.

Gas In the widest sense applied to all aeriform bodies, the most minute particles of which exhibit the tendency to fly apart from each other in all directions. Normally these "gases" are found in that state at ordinary temperature and pressure. They can only be liquefied or solidified by artificial means, either through high pressure or extremely low temperatures.

Gas Air Mixer A device or system of piping and controls which mixes LP-Gas vapor with air to produce a mixed gas of a lower heating value than the LP-Gas. The mixture may replace another fuel gas completely or may be mixed to produce similar characteristics and mixed with the basic fuel gas.

Gas Mask Type of respirator covering entire face, providing eye protection as well as for the nose and mouth; effective against air-containing sprays, dusts, or gases.

Gaseous Diffusion (Plant) A method of isotropic separation based on the fact that gas atoms or molecules with different masses will diffuse through a porous barrier at different rates; requires large plants and enormous amounts of electric power.

Gasolines A mixture of volatile, flammable liquid hydrocarbons used primarily for internal combustion engines; FLASH POINT of approximately −40°C.

Gas Processing Plant A facility designed (1) to achieve the recovery of natural gas liquids from the stream of natural gas which may or may not have been processed through lease separators and field facilities, and (2) to control the quality of the natural gas to be marketed.

Gastrointestinal Syndrome Illness caused by acute exposure to ionizing radiation, resulting in damage to the gastrointestinal system.

Gateway Point at which freight moving from one territory to another is interchanged between railroads.

Gauge of Track The distance between the heads of the rails, measured at a point 5/8 × below top of the rails; standard gauge is 4′ 8 1/2″.

Gauging Nipple A small section of pipe in the top of a tank through which a tank may be measured.

Geiger-Muller Counter (Tube) A radiation detection and measuring instrument, consisting of a gas-filled tube containing electrodes between which there is an electrical voltage but no current flowing. When ionizing radiation passes through, a short intense pulse of current passes from the negative to the positive electrode and is measured. Number of pulses per second measures the intensity of radiation. Also known as a Geiger Counter.

Gelling Agents Chemicals which increase the viscosity of oil, and, in theory, can be applied to an oil slick to reduce its rate of spread over the water surface.

General Emergency (Nuclear reactors) An event involving actual or imminent substantial core degradation (or melting) with potential for loss of containment integrity and with subsequent release of significant radioactivity to the environment. Off-site protective actions may be necessary.

General Service Car RR Box, gondola, or flat car not designed for a specific commodity or shipper, without special equipment.

General Staff (ICS) A group of incident management personnel comprised of but not limited to the Incident Commander, Suppression and Rescue Section Chief, Planning Section Chief, Logistics Section Chief.

Generator Person responsible for the creation of hazardous waste by nature of ownership, management or control.

Genetic Effects of Radiation Radiation effects that can be transferred from parent to offspring; any radiation-caused changes in the genetic material of sex cells.

Geometry The spatial configuration, pattern or relationship of components in an experiment or apparatus. In reactor technology, refers to shape and size of fuel elements, moderator and reflector and their location with respect to each other. In nuclear physics, the arrangement of source and detecting equipment. In counting and scanning, indicates the percentage of radiation leaving a sample that reaches the sensitive volume of a counter.

Glad Hands End of an air hose.

Glasphalt Trade name for a highway paving material wherein recovered ground glass replaces some gravel normally used in asphalt.

Glass A material made from fusion of sand, soda ash and other ingredients. Common glass is impermeable, transparent, sanitary and odorless.

Glove Box A sealed enclosure in which all handling of items inside the box is carried out through long rubber gloves sealed to ports in the walls of the enclosure. Operator places hands and forearms in the gloves from the room side of the box to be physically separated from the gloved box environment, but able to manipulate items inside the box with relative freedom while viewing the operation through a window.

Glycols Any of a class of organic compounds belonging to the alcohol family but having two hydroxyl groups.

Goggles Eye protection device.

Gondola Car Freight car with sides and ends, but without a top covering.

Gooseneck On a drop-frame trailer, the upper level at the front trailer together with the structure connecting it to the lower level. Also used on container chassis to reduce overall height of vehicle.

GPA Gallons per acre.

GPM Gallons per minute.

Grab Iron Steel bar attached to cars and engines as a hand hold.

Gradient (1) Pressure drop. (2) (Brake pipe) Difference in the brake pipe pressure between the front and rear of the train; direct result of leakage or line obstruction.

Grain Door A partition placed across the door of a boxcar to prevent loss of grain by leaking.

Gram The basic unit of weight in the metric system; equal to 1/1000th of a kilogram; approximately 28.5 grams equal one ounce.

Granular Pesticide An active ingredient mixed with or coating small pellets or sand-like material; often used to control soil pests.

Granules Dry, coarse particles of some porous material (clay, corncobs, walnut shells) into which a pesticide is absorbed.

Gravity (Specific) Density expressed as the ratio of the weight of a volume of substance to the weight of an equal volume of another standard substance. In the case of liquids and solids, the standard is water; with natural gas or other gaseous materials, the standard is air.

Gray (Gy) The SI unit for measuring absorbed doses of ionizing radiation. 1 Gy = 1 joule/kg-100 rads.

Gross Ton 2,240 pounds.

Gross Ton-Mile Movement of a ton of transportation equipment and contents a distance of one mile.

Gross Weight The weight of a RR car together with the weight of the entire contents.

Ground Support Unit Leader (ICS) Person responsible for establishing and supervising staging area(s); controlling and dispersing manpower, equipment and apparatus; providing first aid medical care needs of incident personnel; providing for fuel, maintenance and repair of equipment and apparatus.

Groundwater Water present below the soil surface and occupying voids in the porous subsoil; specifically, the porous layer which is completely saturated with water. Upper surface is referred to as the water table.

Groundwater Plume A body of contaminated groundwater originating from a specific source and influenced by the local flow pattern, density and concentration of contaminant, and character of the aquifer.

Groundwater Protection Standard The level of contamination that triggers the need for corrective action measures.

Groundwater Quality The specific chemical, physical, and biological properties of groundwater in a specific area. State and local standards determine its suitability as a drinking water supply.

Growth Regulator A pesticide which increases, decreases, or in some way changes the normal growth or reproduction of a plant or insect.

Guidelines Informal, instructional state or federal agency directives explaining program regulations or policies not addressed clearly by law. Guidelines do not have the force of law.

H

Habitat The sum total of the environmental conditions of a specific place occupied by an organism, a population, or a community.

Hack (Slang) Caboose.

Half-Life A measure of the decay rate of an isotrope to another nuclear form; measured half-lives vary from millionths of a second to billions of years.

Hammermill A type of crusher used to break up waste materials into smaller particles; operates by using rotating and flailing heavy hammers.

Hand Brake (RR) Apparatus utilized to manually apply brakes on a car or locomotive.

Hand Crew (ICS) Predetermined individuals supervised, organized and trained principally for clearing brush as a fire suppression measure.

Hard Facing An extremely hard material, usually crushed tungsten carbide, applied to outside surfaces of tool joints, drill collars, stabilizers, and other oilwell tools to minimize wear.

Hard Water Water containing soluble salts of calcium and magnesium and sometimes iron.

Harm An injury or damage.

Hatch An opening into a tank, usually through the top deck.

Hatch Plan Schematic drawing of location of all cargo on a ship; also, STOWAGE PLAN.

Hay Tank An enclosure filled with hay-like material used to filter oil out of water.

Hazard The chance that injury or harm will occur to persons, plants, animals or property. In pesticides, the risk of danger resulting from a combination of toxicity and exposure.

Hazardous Chemical An explosive, flammable, poisonous, corrosive, reactive, or radioactive chemical requiring special care in handling because of hazards it poses to public health and the environment.

Hazard Class A group of materials designated by the DOT sharing a common major hazardous property, i.e., radioactivity, flammability.

Hazardous Material A substance in a quantity or form posing an unreasonable risk to health, safety and/or property when transported in commerce; a substance which by its nature, containment and reactivity has the capability for inflicting harm during an accidental occurrence; characterized as being toxic, corrosive, flammable, reactive, an irritant, or a strong sensitizer and thereby posing a threat to health and the environment when improperly managed.

Hazardous Materials Categories

Explosive—Any chemical compound, mixture or device, the primary or common purpose of which is to function by explosion, with substantially instantaneous release of gas and heat.

Flammable Liquid—Any liquid having a flash point below 100°F as determined by tests listed in Code of Federal Regulations 49, Sec. 173.115(d).

Combustible Liquid—Any liquid having a flash point above 100°F and below 200°F as determined by tests listed in Code of Federal Regulations 49, Sec. 173.115.

Flammable Gas—Any gas which, in a mixture of 13% or less by volume with air, is flammable at atmospheric pressure, or its flammable range with air at atmospheric pressure is wider than 12% (by volume) regardless of a lower flammability limit.

Nonflammable Gas—Any compressed gas other than a flammable gas.

Flammable Solid—Any solid material, other than an explosive, which is liable to cause fires through friction, retained heat from manufacturing or processing, or which can be ignited readily and when ignited burns so vigorously and persistently as to create a serious transportation hazard.

Oxidizer—A substance that yields oxygen readily to stimulate the combustion of other material.

Organic Peroxide—An organic compound which may be considered a derivative of hydrogen peroxide where one or more of the hydrogen atoms have been replaced by organic radicals, and readily releases oxygen to stimulate the combustion of other materials.

Poison A—A poison gas, extremely dangerous gases or liquids of such nature that a very small amount of the gas or vapor of the liquid, mixed with air is dangerous or lethal to life.

Poison B—Liquids or solids, including gases, semi-solids, and powders other than class a or irritating materials, which are known to be so toxic to man as to afford a hazard to health.

Irritating Material—A liquid or solid substance which upon contact with fire or when exposed to air gives off dangerous or intensely irritating fumes, but not including any Class A poisonous material.

RADIOACTIVE MATERIAL—Also known as Radiological Material, is any material or combination of materials, that spontaneously emit ionizing radiation, and has a specific gravity greater than 0.002 microcuries per gram.

CORROSIVE MATERIAL—Any liquid or solid, including powders, that cause visible destruction of human skin tissue or a liquid that has a severe corrosion rate on steel or aluminum.

ETIOLOGICAL AGENT—A viable microorganism or its toxin which causes or may cause human disease.

CONSUMER COMMODITY—A material packaged or distributed in a form intended and suitable for sale through retail sales agencies for use or consumption by individuals for purposes of personal care or household use. Term also denotes drugs and medicines.

Hazardous Substances Account A California State fund derived from fees paid by persons who submit more than 500 pounds per year of hazardous or extremely hazardous waste to on- or off-site hazardous waste disposal facilities; funding source for the state Superfund program.

Hazardous Waste Waste materials or mixtures of waste which require special handling and disposal because of their potential to damage health and the environment.

Hazardous Waste Constituent Substance contained in waste causing that waste to be listed as hazardous.

Hazardous Waste Control Act A California state law enacted in 1972. The first comprehensive law of this type in the United States, it established the state's hazardous waste management program within the Department of Health Services.

Hazardous Waste Facility Permit Document issued by the Department of Health Services granting the authority to operate a transfer station or a facility that stores, treats or disposes of a hazardous waste.

Hazardous Waste Landfill An environmentally sound disposal facility where hazardous waste can be placed without polluting the environment, not including a land treatment facility, a surface impoundment, or an injection well.

Hazardous Waste Management Systematic control of the collection, source separation, storage, transportation, processing, treatment, recovery, and disposal of hazardous wastes.

Hazardous Waste Site A location where hazardous wastes are stored, treated, incinerated, or otherwise disposed of.

Head The front and rear closure of a tank shell.

Head End Beginning or forward portion of any train.

Head Man Brakeman responsible for work done in connection with the forward section of the train and, when in transit, is stationed in the locomotive; also, HEAD PIN (slang).

Header Identifying portion (beginning) of any list or consist.

Head Wave An area of oil concentration which occurs behind and at some distance from containment booms; area is important to the positioning of mechanical recovery devices and is the region where droplet breakaway boom failure phenomenon is initiated when current flow exceeds critical velocity. *see* DROPLET BREAKAWAY.

Health Facility (ICS) Any facility, place or building organized, maintained and operated for the diagnosis, care and treatment of human illness or injury, physical or mental, including convalescence, rehabilitation and/or pre- and post-natal care, for one or more persons, to which patients are admitted for 24 hours or longer.

Heat A condition of matter caused by the rapid movement of its molecules. Energy has to be applied to the material in sufficient amounts to create the motion, and may be applied by mechanical or chemical means.

Heat (A Connection) To loosen a collar or other threaded connection by striking it with a hammer.

Heat of Fusion *see* LATENT HEAT.

Heat Measurement Temperature; may be calibrated on several scales, most commonly Fahrenheit and Centigrade; measurement of heat quantity is BTU.

Heat of Vaporization *see* LATENT HEAT.

Heat Transfer *see* RADIATION, CONVECTION, CONDUCTION.

Heater (Switch) (RR) A device for melting snow at switches by means of steam, an electric current, gas jets or oil.

Heater Car An insulated box car equipped with heating apparatus for the protection of perishables.

Heating Tube Also called fire tube; a tube device installed inside a tank used to heat the tank contents.

Heavy Equipment Transport (ICS) Any ground vehicle capable of transporting a bulldozer.

Heavy Metals High-density metallic elements generally toxic to plant and animal life in low concentrations (e.g. mercury, chromium, cadmium, arsenic, and lead).

Heavy Water Water containing significantly more than the natural proportion (one in 6,500) of heavy hydrogen (deuterium) atoms to ordinary hydrogen atoms; used as a moderator in some nuclear reactors as it slows neutrons effectively and also has a low cross section for absorption of neutrons.

Hectare A metric area measurement equal to 10,000 square meters or approximately 2.47 acres.

Helibase, Helispot A location within an incident area where helicopters may be landed, serviced and loaded.

Hemispherical Head Used on MC-331 high pressure tanks, head is in shape of a half-sphere.

Hepa Filter *see* ABSOLUTE FILTER.

Herbicide Pesticide used to control plant life.

High Iron (Slang) Mainline or high speed track of a RR system.

High Rail The outer or elevated rail of a curved track.

High Side Gondola Car Gondola car with sides and ends over 36″.

High Expansion Foam Detergent-base foam with low water content which expands in ratio of 1000 to 1.

Highball (RR slang) Signal to proceed at maximum authorized speed.

Hitch A connecting device at rear of a vehicle used to pull a full trailer with provision for easy coupling.

Hog (slang) Locomotive.

Hog Head (slang) Locomotive engineer.

Hogger (slang) Locomotive engineer.

Hog Law (slang) Federal statute providing that all train and engine crews must be relieved of duty after 14 hours of continuous service (to be changed to 12 hours at a later date).

Hold Track RR track where cars are held awaiting disposition.

Hole (slang) An area of track enabling one train to pass another.

Home A location where a RR car is on the tracks of its owner.

Home Car A car on the tracks of its owner.

Home Junction (RR) A junction with the home road.

Home Road The owning road of a railroad car.

Home Route Return route of a foreign empty RR car to the owning road.

Home Scrap Scrap utilized within the plant where it originates.

Home Signal (RR) A fixed signal at entrance to an interlocking to govern trains or engines entering and using that block.

Hook (slang) A crane used in wreck train service. Also: "big hook," "wrecker."

Hopper (1) An open-top RR car with one or more pockets opening on the underside of the car to permit quick unloading of bulk commodities. (2) Sloping panels at the bottom of a tank which direct dry bulk solids to the outlet piping.

Horsepower A measure of power; one horsepower is equivalent to a force that will raise 33,000 pounds one foot in one minute.

Hose Tube A housing used on tank and bulk commodity trailers for the storage of cargo-handling hoses; also: hose troughs.

Hostler A RR fireman who operates light engines in designated enginehouse territory and works under the direction of the engine house foreman.

Hostler's Control A simplified throttle provided to move the "B" unit of a diesel locomotive not equipped with a regular engineer's control.

Hot Box An overheated journal caused by excessive friction between bearing and journal, lack of lubricant or foreign matter.

Hot Box Detector A wayside infrared sensing instrument for determining journal temperature.

Hot Cell A heavily shielded enclosure in which radioactive materials can be handled remotely with manipulators and viewed through shielding windows to limit danger to operating personnel.

Hot Spot Radioactive area.

Hump An incline in a RR yard over which cars are uncoupled and allowed to roll free into a classification yard.

Hy-Cube Car A box car of approximately 85′ length and 10,000 cubic foot capacity designed for hauling automobile body stampings and other low density freight.

Hydrapulper Trade name for a large mechanical device used primarily in paper industry to pulp waste paper or wood chips and separate foreign matter; suspends finely divided cellulose fibers in water.

Hydraulic Dispersion Shoreline cleanup technique utilizing a water stream at either high or low pressure to remove stranded oil; most suited for removal of oil from coarse sediments, rocks, and man-made structures.

Hydration Process in which particles go into a water solution and become surrounded by a sheath of water molecules.

Hydrocarbon A chemical containing only carbon and hydrogen atoms. Crude oil is a mixture largely of hydrocarbons; methane (natural gas) is simplest with each molecule containing one atom of carbon and four atoms of hydrogen.

Hydrogen The lightest element, No. 1 in the atomic series; has two natural isotopes of atomic weights 1 and 2; first is ordinary (or light) hydrogen, second is deuterium (heavy) hydrogen. A third isotope, tritium, atomic weight 3, is a radioactive form produced in reactors by bombarding lithium 6 with neutrons.

Hydrogen Bomb Nuclear weapon deriving its energy largely from fusion.

Hydrogen Ion Concentration Measure of acidity or alkalinity, expressed in terms of the pH of the solution. See PH.

Hydrology Study of the properties, distribution and flow of water on or within the earth.

Hydrolysis Hazardous waste chemical treatment method wherein chemical compounds are decomposed by a reaction with water; agents such as alkaline solutions as well as high temperatures and pressures are often used to promote desired reaction.

Hydrophilic "Water loving"; referring to molecules that associate with water and are readily wet by water.

Hydrophobic "Water fearing"; molecules that are poorly soluble in water, are water repellant or not made wet by water.

Hydrophobic Agent Chemical having the ability to resist wetting by water; occasionally used in treatment of synthetic sorbents to decrease the amount of water absorbed hence increasing the volume of oil they can absorb before becoming saturated.

Hydrophyte A plant growing in water or soil too waterlogged for most plants to survive.

Hygroscopic Ability of a substance to absorb moisture from the air.

Hypergolic Any substance that spontaneously ignites upon contact with another. Many hypergolics are used as rocket fuels.

I

ICC—Interstate Commerce Commission Independent federal government agency in the Executive Branch (not affiliated with DOT) charged with administering acts of Congress affecting rates and routes for transport of interstate commerce.

ICC Cylinder *see* CYLINDER.

Ice Bunker (Refrigerator car) Compartment where ice is placed.

Icing Charge Made for icing perishable freight.

ID Inside diameter.

Identification Code The individual number assigned each hazardous waste generator, transporter, and treatment, storage or disposal facility by state or federal regulatory agencies.

Identification Lamps On the rear of a vehicle, a cluster of three red lamps mounted as close as practicable to the top of the vehicle at the same height, one on the vertical centerline, one on each side of the vertical centerline with lamp centers spaced not less than 6 inches nor more than 12 inches apart.

Idler Car An unloaded flat car used to protect overhanging loads or used between carrying cars loaded with long material.

IFR Instrument Flight Rules; apply when visibility is less than 3 miles and ceiling is below 1,000 feet.

Ignitable Waste A liquid with a flash point less than 60°C (140°F), a waste which is an oxidizer, or ignitable compressed gas or non-liquid which is liable to cause fires through friction, absorption of moisture, spontaneous chemical changes or when ignited burns so vigorously and persistently as to create a hazard.

Ignition Temperature The temperature to which a substance must be heated in order to initiate self-sustained combustion (burning).

IGR Insect growth regulator.

Illegal Residue A residue in excess of a pre-established and government-enforced safe level.

ILS Instrument Landing System; radio aid to navigation to assist aircraft in landing.

Immiscible Not capable of mixing or being mixed (as oil and water).

Immobilization Threshold The minimal amount of a substance causing cessation of movement in a test organism when applied in a particular manner.

Immune Exempt from or protected against; a state of not being affected by a disease or poison.

Impact Register An appliance placed in a RR car with shipment which is both a time clock and a measuring device to record amount of shock the car received enroute.

Impermeability As applied to soil or subsoil, the degree to which fluids, particularly water, cannot penetrate in measureable quantities.

Impermeable Cannot be penetrated by gases or liquids. Semi-permeable means that some substances can penetrate and others cannot.

Import To receive goods from a foreign country.

Improvised Booms Booms constructed from readily available materials such as railroad ties, logs and telephone poles. Used as containment measures for hazardous materials spills.

Inactive Not involved in action; in pesticides, not reacting with anything.

Inactive Facility The EPA designation for a treatment, storage, or disposal facility that has not accepted hazardous waste since November 19, 1980.

Inactive Portion A portion of a hazardous waste management facility which has not operates since November 19, 1980, but which is not yet a closed portion (no longer accepts waste to that area).

In-Bond Shipment An import or export shipment which has not been cleared by federal customs officials.

Inbound Train Train arriving at a yard or terminal.

In the Clear (RR) When a car or train has passed over a switch and frog so far that another car or train can apss without collision.

In the Hole (slang) On a siding.

In-Plant Waste Waste generated in manufacturing processes; such might be recovered through internal recycling or through a salvage dealer.

Incentives Measures providing benefits to communities above and beyond costs associated with hazardous waste management facilities. Also refers to certain measures (such as low interest loans, etc.) taken by government to stimulate development and implementation of improved hazardous waste management technologies.

Incident An occurrence; involving radioactive materials, an event resulting in the loss of control of radioactive materials and involves an immediate or likely hazard to life, health, or property.

Incident Action Plan (ICS) A preplan containing general control objectives reflecting the overall incident strategy and specific suppression and rescue action plans for the next operational period.

Incident Commander The individual having the responsibility for total operations at a hazardous materials emergency.

Incident Command Post Command Post.

Incineration The application of high temperatures (800°–3000°F) to break down organic wastes into simpler forms and to reduce the volume of waste needing disposal. Energy can be recovered from incineration heat.

Incineration Technologies The processes by which waste volume is reduced by combustion in a controlled manner, primary purpose to thermally break down hazardous waste.

LIQUID INJECTION SYSTEMS—Brick-lined chambers or beds of inert granular material into which solid and liquid wastes are injected for incineration.

ROTARY KILNS—Large, refractory cylinders capable of burning virtually any solid or liquid organic waste.

CEMENT KILNS—Used in the production of cement and in which organic wastes can be burned at temperatures ranging from 2600° to 3000°F.

Incinerator A plant designed to reduce waste volume by combustion. For pesticides, a special high-heat furnace which reduces everything to non-toxic ashes and vapors; used for disposing of highly toxic pesticides.

Incipient Fires Fires in the beginning stages.

Incompatible Not agreeable; when two or more pesticides are mixed and effectiveness of one or more is decreased; not capable of being mixed or used together. Also, when two or more pesticides are mixed and application causes unintended injury to plants or animals.

Incompatible Waste (1) A hazardous waste unsuitable for placement within a specific portion of a landfill because it may cause containment material to corrode or decay or, when combined with other wastes, might produce heat, pressure, fire, explosion, violent reaction, toxic dusts, mists, fumes, or gases. (2) Hazardous wastes which, if mixed, would become more hazardous than either waste individually.

Incompatibilities A list of substances and conditions which can produce undesirable and potentially dangerous effects when stored, packaged, mixed or handled together.

Independent Brake Valve (Air Brake) Device to operate the locomotive brakes independently of the train brakes.

Indication Information conveyed by the aspect of a RR signal.

Individual Container Cargo container, such as a box or drum, used to transport materials in small quantities.

Induce Vomiting Make a person or animal throw up stomach contents.

Induced Radioactivity Radioactivity created when substances are bombarded with neutrons, as from a nuclear explosion or in a reactor, or with charged particles produced by accelerators.

Industrial Plant (LP-Gas) A facility utilizing gas incidental to plant operations with LP-Gas storage of 2,000 gallons water capacity or more, and which receives gas in tank car, truck transport or truck lots. Normally LP—Gas is used through piping systems in the plant, but may also be used to fill small containers such as for engine fuel on industrial (i.e., forklift) trucks. Only plant employees have access to these filling facilities so are not considered to be distributing points. *see* DISTRIBUTION POINT.

Industrial Scrap Waste generated during a manufacturing operation.

Industrial Waste Waste materials generally discarded from industrial operations or derived from manufacturing processes; may be liquid, sludge or solid wastes and need not be hazardous.

Inert Any substance that will not react chemically; inactive.

Inert Ingredient Inactive ingredient; substance(s) in a pesticide product which has no killing or controlling action.

Infectious Waste Waste containing pathogens; may consist of tissues, organs, body parts, blood, and body fluids removed during surgery.

Infiltration Flow of a fluid into a substance through pores or small openings; commonly used to denote flow of water into soil material.

Inflammable Liquids Liquids emitting vapors which become combustible at a certain temperature.

Ingestion The taking in of toxic materials through the mouth; to eat, swallow, drink, or in some other way take into the digestive system.

Ingestion Pathway Possible route by which radioactive material is introduced into the environment and subsequently ingested by living beings; principal exposure would be contaminated water or foods, especially near a nuclear facility.

Ingredient Statement The listing on the label of a pesticide container giving the name and amount of each active ingredient and the total of inert ingredients in the product.

Inhalation The process of absorbing toxic materials by breathing through the nose or mouth, taking air into the lungs, breathing in.

Inhibitor Substance added to an unstable material that prevents its entrance into a violent reaction; a stabilizer.

Initial Attack (ICS) Resources committed at the beginning of an incident.

Initial Carrier Railroad on which a shipment originates.

Initial Point Location at which a shipment originates.

Injection Subsurface placement of a fluid or waste.

Injection Well A well into which fluids are forced.

Injury Disruption of the intended functional continuity of an animate or inanimate thing; may range from negligible to fatal.

Injury Mechanism Process by which a material interacts to injure.

Inner Liner Continuous layer of material placed inside a tank or other container to protect the construction materials of that container from the contents.

Inorganic Substance which does not contain carbon.

Inorganic Compounds Chemical compounds not containing carbon.

Inorganic Pesticides Pesticides which do not contain carbon; examples, Bordeaux mixture, copper sulfate, sodium arsenite, sulfuric acid, and salt.

Insect Growth Regulator A synthetic organic pesticide which mimics insect hormonal action so that the exposed insect cannot complete its normal development cycles and dies without becoming an adult.

Insecticide Chemical used to control insect pests.

Instrument Landing Landing an aircraft by reliance on instrument data with no external vision.

Insulated Rail Joint A rail joint arresting the flow of electric current from rail to rail, as at the end of a track circuit, by means of nonconductors separating rail ends and other metal parts.

Integrated Control Use of more than one method of pest control, including cultural practices, natural enemies, and selective pesticides.

Intensity Amount of radiation passing through a unit area at a given time; measured in terms of roentgens by radiological survey instruments.

Inter- Prefix synonymous with between.

Interchange Exchange of cars between railroads at specified junction points.

Interchange Point Location where RR cars are transferred from one road to another.

Interchange Track Track on which freight is delivered by one railroad to another.

Interim Authorization Conditional permission from EPA which enables a state to operate its own hazardous waste management program.

Interim Status A period of time when hazardous waste storage, treatment facilities and transporters could continue to operate under a special set of regulations until the appropriate permit or license application is or was approved by DER.

Interline Between one or more railroads.

Interline Freight Freight moving from point of origin to destination over lines of two or more railroads.

Interline Waybill Document covering movement of freight over two or more railroads.

Interlocking An arrangement of RR signals and switches set in such a way that their movements must succeed each other in a predetermined order so that a clear indication cannot be given simultaneously on conflicting routes. Found at a crossing of two railroads, a drawbridge, junction, or entering or leaving a terminal or yard.

Interlocking Limits Tracks between the extreme opposing home signals of an interlocking.

Intermediate Carrier A railroad over which a shipment moves but on which neither the point of origin nor destination is located.

Internal Radiation Radioactivity originating within the body. Usually deposited in the body by inhalation or ingestion of a radioactive contaminant.

Internal Valve A primary shutoff valve for containers which has adequate means of actuation and is constructed in such a manner that its seat is inside the container; damage to parts exterior to the container or mating flange will not prevent effective seating of the valve.

Interstate Commerce Act Act of Congress regulating practices, rates and rules of transportation lines engaged in handling interstate traffic.

Interstate Commerce Commission Regulating body of the United States Government having jurisdiction over transportation matters.

Interstate Traffic Traffic moving from a point in one state to a point in another state or between points in the same state, but passing within or through another state enroute.

Interval The time between two occurrences; time between two pesticide applications; also the period between the final pesticide application and harvest.

Intra- Prefix synonymous with within.

Intraplant Switching Moving of RR cars from one place to another within the yards of a plant or industry.

Intrastate Traffic Traffic having origin, destination, and entire transportation within the same state.

Inverse Square Law Mathematical relationship stating that radiation intensity is inversely proportional to the square of the distance from the source; as the distance doubles, the intensity decreases by one-fourth.

Invert Emulsifier An adjuvant allowing water to remain suspended in oil rather than settling out; reverse of the usual emulsifier which allows suspension of oil in water.

Ion A small particle with a net positive or negative electrical charge.

Ion-Pair Produced when one or more electrons are separated from an atom. The electron is one-half of the ion pair, the nucleus the other.

Ionization Process whereby neutral atoms or groups of atoms become electrically charged, either positively or negatively, by the loss or grain of electrons; process of producing ion-pairs.

Ionization Chamber An instrument detecting ionizing radiation by measuring the electrical current that flows when radiation ionizes gas in a chamber, making the gas a conductor of the electricity.

Ionizing Radiation Energy which interacts with matter by forming ion-pairs.

Iron (slang) RR switch.

Irradiation Exposure to radiation, as in a nuclear reactor or fallout field.

Irritating Materials Liquids or solid substances which give off dangerous or intensely irritating fumes upon contact with fire or when exposed to air.

ISO International Organization for Standardization. An international standards-writing body headquartered in Geneva, Switzerland, composed of national standards associations from some 55 countries. All member countries are given equal status and are entitled to one vote regardless of size or economic development. Technical work is carried on in committees.

Isolated Source A single, usually small, portable, and self-contained source of ionizing radiation, often located in an area where no other sources are present. Examples: smoke detectors, radiopharmaceuticals, and radiography cameras.

Isolation Perimeter Area around any hazardous materials incident within which only necessary personnel with full protective gear are allowed.

Isomer One or more substances with the same composition but with different properties.

Isotonic A solution having the same osmotic pressure as blood.

Isotope One of two or more atoms with the same atomic number (the same chemical element) but with different atomic weights; isotopes usually have very nearly the same chemical properties but somewhat different physical properties.

Isotopic Enrichment A process by which the relative abundances of the isotopes of a given element are altered, thus producing a form of the element that has been enriched in one particular isotope. Example: enriching natural uranium in the uranium 235 isotope.

IV Intravenous.

J

Jacket Metal cover which protects the tank insulation.

Jackknife Condition of truck tractor semitrailer combination when their relative positions to each other form an angle of 90 degrees (or less) about the trailer kingpin.

Jet Fuel A kerosene or kerosene-based fuel used to power jet aircraft combustion engines.

Jib (slang) Derrick or crane; boom.

Joint Agent Person having the authority to transact business for two or more railroads.

Joint Bar (RR) SPLICE BAR; used in pairs, one on each side of the rail to fasten together the ends of rails; designed to fit the space between head and base closely and held in place by track bolts and suitable accessory equipment.

Jointly Liable When two or more persons or companies share legal responsibility for negligence.

Joule In physics, a unit of work or energy equivalent to 1 watt-second.

Journal End of the axle which moves in the bearings.

Journal Bearing A combination of rollers and races or a block of metal in contact with the journal on which the load rests. In RR car construction the term when unqualified means a car axle journal bearing. *see* ROLLER BEARING.

Journal Box Metal housing enclosing the journal of a car axle, the journal bearing and wedge and which holds the oil and lubricating device for lubricating the journal.

Journal Brass Journal bearing.

Judicial Review A consideration by the courts concerning administrative agency decisions and actions.

Jumper Flexible cable composed of one or more conductors of electric current used to connect electrically the controller circuits between cars or locomotives.

Junction Point (1) Point at which a branch line RR track connects with a main line track. (2) Location at which two or more railroads interchange cars over connecting tracks.

Jurisdictional Agency (ICS) The agency having fire protection responsibility for a specific geographical area.

K

Kaolin A white clay derived from rock composed chiefly of feldspar and with hydrous aluminum silicate as the main chemical ingredient.

Keeper (RR slang) Latch.

Kerosene Flammable oil characterized by a relatively low viscosity and flash point near 55°C; lies between gasolines and fuel oils in terms of major physical properties.

Kg Kilogram.

Kick (RR slang) Act of pushing one or more cars at a speed sufficient to allow free forward movement into selected tracks when uncoupled and engine power reduced.

Kill a Well To overcome pressure in an oil well by use of mud or water so surface connections may be removed.

Kilo Prefix meaning 1,000.

Kilogram Metric weight measurement; (kg) = 1000 grams or approximately 2.2 pounds.

Kiloton Energy The energy of a nuclear explosion equivalent to that of an explosion of 1000 tons of TNT.

Kinetic Energy Energy that results from motion of a mass.

Kingpin Attaching pin on a semitrailer that mates with pivots within the lower coupler of a truck tractor or converter dolly while coupling the two units together.

Knockout A kind of tank or filter used to separate oil and water.

Known Damage Damage discovered before or at the time of delivery of a shipment.

Knuckle (Coupler) The rotating coupling hook by which coupling is effected when the knuckle is locked by the catch or lock.

Knuckle Pin (Coupler) Pin holding the knuckle in the jaws of the coupler; pivot pin.

Knuckle Thrower Device which throws the knuckle of a car coupler open when the uncoupling lever is operated.

L

Label Written or printed matter accompanying an article or substance to furnish identification; in pesticides, contains technical information.

Labeling Technical information about a substance in form of printed material provided by the manufacturer or its agent, including the label, flyers, handouts, leaflets, and brochures.

Labels (1) Four-inch square diamond markers required on individual shipping containers that are smaller than 640 cu ft. (2) Diamond-shaped warning signs which are required to be placed on all containers of radioactive material. *see* PLACARD.

Ladder (RR) Main track of a yard from which individual tracks lead. *see* LEAD TRACK.

Lading Freight or cargo making up a shipment.

Lagoon Shallow pond used for the temporary or permanent storage of liquid; sunlight, bacterial action, and oxygen interact to restore wastewater to a reasonable state of purity.

Land Burial Disposal of wastes into land (landfills, surface impoundments, permanent waste piles, and underground injection wells); used for hazardous substances requiring permanent storage.

Land Disposal Practice of disposing of liquid and solid hazardous wastes in earthen pits. In California, the State Water Resources Control Board licenses two classes of land disposal facilities which can accept hazardous wastes:

CLASS I—Sites which cannot overlie usable ground water, except in "extreme cases," and which may receive all classes of hazardous wastes except PCBs and radioactive wastes.

CLASS II-1—Sites which may overlie or be adjacent to usable ground water, but must protect ground water by natural site characteristics or site modifications.

Land Disposal Methods of Dealing with Hazardous Waste Materials:

1. Landfill, surface impoundment, waste piles, deepwell injection, land spreading, and co-burial with municipal garbage.

2. Treatment such as neutralization and evaporation ponds and land farming where residues are hazardous and are not removed for subsequent processing or disposal within one year.

3. Storage such as waste piles and surface impoundments for longer than one year.

Land Disposal Restrictions A California state program administered by the Department of Health Services designed to progressively ban the land disposal of certain hazardous wastes.

Landfill (Sanitary landfill) Facility for disposal of solid or hazardous waste involving burial in an excavated area or natural depression.

Landfill Cell Compacted solid waste enclosed by soil or cover material within a landfill.

Landing The whole procedure of maneuvering an aircraft to a stop.

Landing Depth The depth to which the lower end of oilwell casing extends in the hole when installed.

Land Treatment Facility A location at which hazardous waste is applied or incorporated into the soil surface; such facilities are disposal facilities if the waste will remain after closure.

Larva A young growing insect in the stage after hatching from the egg and before becoming a pupa; worm- or grub-like, usually not resembling the adult stage. Many insects cause most or all of their damage as larvae.

Larvacide A pesticide used to kill the larvae of insects.

Latent Heat The amount of heat absorbed or given off by a substance as it passes from a liquid to a gas (HEAT OF VAPORIZATION), or a solid to a liquid state (HEAT OF FUSION); measured in terms of the number of BTUs required to complete the phase.

Latent Period A symptom-free period in acute radiation syndrome lasting from 3 to 21 days; follows prodromal period.

Lateral Motion (RR) Motion which takes place, crosswise of the track, of all car parts except the wheels and axles; results from the flexibility which must be provided in truck structure to permit safe negotiation of track curves.

Lawful Rate A rate published in conformity with the provisions of the regulatory law and which does not violate any other provisions of such law.

LC50 Concentration of an active ingredient in the air which, when inhaled, kills half of the test animals exposed to it; expression of a compound's toxicity when present in the air as a gas, vapor, dust, or mist; generally expressed in ppm when a gas or vapor, and in micrograms per liter when a dust or mist; often used as the measure of ACUTE INHALATION TOXICITY. The lower the LC50 number value the more poisonous the pesticide.

LD50 Dosage or amount of an active ingredient which, when taken by mouth or absorbed by the skin, kills half of the test animals exposed; an expression used to measure ACUTE ORAL or ACUTE DERMAL TOXICITY.

LD100 The dose of an active ingredient taken by mouth or absorbed by the skin which is expected to cause death in 100% of the test animals so exposed.

Leachate A liquid that has percolated through solid waste or other material and carries decomposed waste, bacteria and other noxious and potentially harmful materials which drain from landfills in solution or suspension and must be collected and treated so as not to contaminate water supplies. Ground water contaminated by leachate is difficult to clean up and may take decades for the contaminants to be flushed out naturally.

Leaching Movement of a substance downward or out of the soil as the result of water movement.

Lead Agency The public organization having principal responsibility for approving a project or executing a plan.

Lead Track (RR) An extended track connecting either end of a yard with main track.

Leaks and Spills Procedures Utilizes Hazardous Materials Storage and Transportation guidelines to provide instructions for dealing with air, water and land spills and contamination by infectious agents. Spells out DOT firefighting recommendations and evacuation instructions; gives Hazardous Materials Table and Optional Hazardous Materials Table classification for the labeling, transportation and storage of hazardous materials.

Legal Residue A remainder of a substance that is within a safe level according to the regulations.

Less than Carload (RR) Quantity of freight less than that required for the application of a carload rate.

Less than Carload Rate (RR) Rate applicable to a less than carload shipment.

Lethal Causing or capable of causing death; deadly; fatal.

Lethal Concentration Amount of toxic substance in air which will likely cause death if inhaled.

Lethal Dosage (1) Amount of a toxic substance which is likely to cause death when ingested. (2) Dose of ionizing radiation sufficient to cause death; media lethal dose (MLD or LD50) is amount required to kill within a specified period of time (usually 30 days) half of the organisms exposed; the LD50/30 for people is about 400–450 roentgens.

Lethal Time (LT) The time required for a defined dose of toxicant to produce a given mortality level in a test organism.

Leukocytes White blood cells.

Liable Legally responsible.

Liability Something that someone is legally responsible for.

Liaison Officer (ICS) Person responsible for interaction with assisting and cooperating agencies.

Licensing The issuing of a certificate of permission from a constituted authority to conduct a service, profession, or business.

Lift Pads Protective devices affixed to lower side rails of a TOFC trailer to provide distribution of stresses at crossmembers and lower rail area during trailer lifting and handling operations.

Light Ends Petroleum products which have low flash points and high vapor pressures; generally referring to liquefied petroleum gas. The light ends are the first compounds recovered from crude oil during the fractional distillation process and are also the first fractions of spilled oil to be lost through evaporation.

Light Engine RR engine moving without caboose or cars attached.

Light Weight Weight of an empty freight car.

Lighter (RR) A flat-bottomed boat used in inland waterways.

Lighterage Limits Area within which freight is handled by barges under certain specific charges, rules and regulations.

Lightering Hauling of freight on barges.

Limited Advanced Life Support (LALS) (EMS) Special service designed to provide prehospital emergency medical care with certain limitations.

Limited Speed (RR) Not exceeding 45 miles per hour.

Lin-Log Display Combination meter for continuous display of radiation readings over a wide range.

Line-Haul Movement of RR freight over the tracks of a railroad from one town to another; not a switching service.

Line-Haul Switching Movement of RR cars within the yard or switching limits of a station, preceding or following a line-haul move.

Liner Continuous layer of natural or synthetic materials beneath or on the sides of a storage or treatment device which severely restricts downward or lateral escape of hazardous waste or leachate.

Liquefied Gases Gases which are liquefied at normal temperatures by placing them under pressure.

Liquefied Petroleum Gas Material having a vapor pressure not exceeding that allowed for commercial propane composed predomininantly of the following hydrocarbons, either by themselves or as mixtures: propane, propylene, butane (normal or iso-butane), and butylenes.

Liquid A substance in fluid form which will take the shape of any container in which it is placed.

Liquid Organics Recovery The chemical or physical processing of certain hazardous wastes to separate contaminants from usable materials; resulting products can be reintroduced to the marketplace.

Liter Metric volume measurement equal to one cubic decimeter or a little more than one quart liquid measure.

Lithium Fluoride Material used in personnel dosimeters which has the capability of storing energy when exposed to ionizing radiation and of releasing light later when heated to 200°F.

Litter Solid waste discarded outside established collection, disposal system.

Livestock Car Special RR car for handling of livestock.

Livestock Waybill Special document used for livestock shipments containing feeding and watering instructions for the stock.

Load Limit Maximum load in pounds which a RR car is designed to carry.

Load, Loading To fill.

Local Involving only a restricted part of the organism; a limited district; referring to a specific community or area.

Local Agency (ICS) Any agency having jurisdictional responsibility for all or any part of an incident.

Local Assessment Committee A review group created by a host or abutting community to analyze a proposed hazardous waste management facility. In some states such committees have authority to negotiate with the facility proponent on behalf of the community concerning the conditions under which a hazardous waste management facility may be built.

Local EMS Agency The organization having primary responsibility for administration of emergency medical services in an area.

Local Rate (RR) Rate applying between stations located on the same railroad.

Local Veto Authority The ability of cities and counties to unilaterally reject proposed facilities by denying local land use approval for hazardous waste management facility siting.

Local Waybill A document covering movement of freight over a single railroad.

Locomotive Engine on a train.

Log A systematic recording of data.

Logistics Section Chief (ICS) Individual responsible at an emergency incident for management of units providing personnel, apparatus, equipment, facilities, and personal needs.

Long and Short Clause Fourth Section of the Interstate Commerce Act; prohibits railroads from charging more for a shorter than for a longer haul over the same route, except by special permission of the ICC.

Long-Term Care The post-closure monitoring and maintenance of a hazardous waste management facility in a manner that protects public health and the environment.

Long-Term Exposure Exposure (to radiation) lasting more than four days.

Long Ton 2,240 pounds; Gross Ton.

Look-Out Caboose, cupola.

Lorry Small four-wheel push car used in railroad construction and maintenance work for moving rails, ties, etc.

Low Concentrate Solution(s) Mixture containing a small amount of an active ingredient in a highly refined oil; usually utilized as stock sprays and space sprays and for use in aerosol generators.

Low Population Zone Area containing residents immediately surrounding an exclusion area when the number and density of that population is such that there is a reasonable probability that appropriate protective measures could be taken in their behalf in the event of a serious accident.

Low Rail Inner rail of a curve which is maintained at grade while the opposite or outer rail is elevated.

Low-Side Gondola Gondola RR car with sides and ends 36 inches high or less.

Low Specific Activity Legal classification for radioactive shipments which allows the use of the radioactive LSA label in place of the radioactive I, II, or III label on sole-use vehicles; signifies (1) natural radioactive materials such as ores, (2) materials with a high specific activity which are dispersed throughout a large volume of inactive material, or (3) non-radioactive items having a low concentration of surface contaminations. *see* ACTIVITY.

LPG Liquefied Petroleum Gas.

LP—Gas Service Station Facility open to the public which consists of LP-Gas storage containers, piping and pertinent equipment, including pumps and dispensing devices, and any buildings in which LP-Gas is stored and dispensed into engine fuel containers of highway vehicles. *see* DISTRIBUTING POINT.

LSA Low Specific Activity.

Lubricating Oils Petroleum-based oils used to reduce friction and wear between solid surfaces such as moving machine parts and internal combustion engine components.

Lunette Metal ring at end of tow bar for attachment of the two bar to the towing vehicle.

M

Magnaflux Trade name for equipment and process used for detecting cracks in iron or steel. A magnetic field is set up in the part to be inspected and a powder or paste of magnetic particles is applied; those particles arrange themselves around discontinuities in the metal, revealing defects.

Magnetic Gauge *see* FLOAT GAUGE.

Magnetic Separator Equipment used to remove magnetic materials from other materials, usually consisting of a belt, drum or pulley and magnet.

Main Deck The principal deck extending from front to back of a ship or offshore drilling rig.

Main Iron (RR) Main track.

Main Line A large diameter pipeline between distant points; trunk line.

Main Reservoir On RR air brakes, the tank on the engine for storing the main air supply, to distinguish from the auxiliary reservoirs under each car.

Maintenance All actions taken to retain material in a serviceable condition to preserve serviceability; includes inspection, testing, servicing, classification as to serviceability, repair, rebuilding, and reclamation; routine, recurring work required to keep a facility in such condition to be continuously utilized in its original or designed capacity and efficiency.

Main Track A designated track upon which trains are operated by timetable, train order, or both, or the use of which is governed by block sygnals.

Make a Connection To attach a joint of oilwell drill pipe onto the drill stem suspended in the wellbore to permit depening of the wellbore.

Make a Joint (RR slang) (1) Couple RR cars together. (2) To screw a length of oilwell pipe into another length of pipe.

Management (Hazardous Waste) A program for controlling the generation, storage, collection, transportation, treatment, use, conversion or disposal of hazardous wastes; includes administrative, financial, legal and planning activities as well as operational aspects of hazardous waste handling, disposal and resource recovery systems.

Man-Machine System All of the personnel, equipment, and other resources capable of functioning during an emergency to return a community to normal conditions with the least amount of injury or damages.

Manhole Openings usually equipped with removable, lockable covers and large enough to admit a man into a tank trailer or dry bulk trailer.

Manifest A description of the contents of a shipment.

Manifest Illness Peak illness period of acute radiation syndrome lasting from 21 to 45 days; follows the latent period.

Manifest Train Scheduled freight train.

Manifold Junction of a number of discharge pipelines to a common outlet.

Manipulators Mechanical devices used for safe handling of radioactive materials; frequently remotely operated from behind a protective shield.

Manometer A U-shaped glass tubing containing a liquid (usually water or mercury) used to measure the pressure of gases or liquids. When pressure is applied, liquid level in one arm rises while the level in the other drops. A set of calibrated markings beside one of the arms permits a pressure reading to be taken, usually in inches or millimeters.

Manpower and Training Component (EMS) Involves resources and arrangements for initial and continuing education and training of all personnel providing EMS including periodic proficiency testing, appropriate credentialing and adequate career mobility and self-fulfillment opportunities.

Manual Block Signal System RR signal system wherein use of each block is controlled by hand.

Manual Recovery Recovery of oil from contaminated areas by cleanup work force using buckets and shovels; extremely labor intensive.

Marked Capacity RR car load capacity as designated on that car.

Marker Front and rear signals of a train (flags or lamps).

Marker Beacon Radio beacon which radiates vertically a distinctive pattern for providing position information to aircraft.

Marking A sign, inscription, symbol or visible impression on an article, like a container.

Marl A semisolid or unconsolidated clay, silt, or sand.

Mass The property of a body that is a measurement of its inertia.

Mass Casualty (Disaster) (EMS) An emergency involving so large a number of casualties and/or a response environment which has major barriers to the delivery of care that require the response of multiple public agencies and delayed transport of the most seriously injured.

Mass Number Total number of protons and neutrons in an atom.

Mast A portable oil derrick capable of being erected as a unit as distinguished from a stand derrick which cannot be raised to a working position as a unit.

MAST (1) Medical Anti-Shock Trousers. (2) (EMS) Military Assistance to Safety in Traffic; a program utilizing military helicopters and medical corpsmen to augment the existing EMS system.

Master Gate A large valve used to shut in an oil well.

Master Valve A large valve on the Christmas tree (*see* VENT) used to control the flow of oil and gas from a well.

Material Substance; in pesticides often used to mean a formulation, chemical, active ingredient or additive ingredient.

Materials Market Combined commercial interests that buy substances; in the recycling industry the market for products made from recycled materials determines the economic feasibility of recycling and resource recovery.

Matter That which has weight (mass) and occupies space; divided into three classes according to condition and state: solid, liquid, or gas. All changes matter exhibits are classified as physical or chemical; if composition is not altered in the process the phenomena is said to be physical; if a change takes place it is referred to as chemical reaction. No destruction of matter is possible, only change to a different state. Often identified by its properties such as density, conductivity, solubility, melting point, boiling point, hardness, color, etc.

Maximum Dosage Largest amount of a pesticide that can be used safely without excess residues or damage occurring to whatever is being treated.

Maximum Rate Highest RR rate that may be charged.

Meat Rack (RR refrigerator car) Supports near the ceiling from which meat is suspended; beef rail.

Mechanical Agitation Stirring, paddling, or swirling action of a device which keeps a pesticide and any additives thoroughly mixed in the spray tank.

Mechanical Refrigerator Car A RR car equipped with a diesel powered refrigerating unit under thermostatic control.

Mediation A voluntary negotiation process wherein a neutral mediator assists the parties in a dispute to reach a mutual agreement.

Medical Emergency (EMS) A situation with a real or perceived need for immediate medical care based on injury or other unforeseen acute physical or mental disorder.

Medical Mutual Aid (MMA) (EMS) The provision of medical assistance to a requesting county by another government jurisdiction when that affected county does not have sufficient medical resources to respond effectively to a disaster.

Medical Officer Individual in the Command Post responsible for the observation and medical treatment of those working at the site of a hazardous materials incident.

Medium Speed Not exceeding 30 miles per hour.

Mega- Prefix meaning one million (1,000,000).

MHz Megahertz; an identification of radio frequencies.

Melting Point Temperature at which a solid is changed to a liquid.

Membrane The impermeable layer of a sanitary landfill liner.

Memorandum of Agreement A written accord between administrative agencies which clarifies or establishes joint procedures or authorities necessary to administer a program.

Memo Waybill Document used when agent does not have sufficient information to determine freight charges; contains adequate information to properly handle car.

Memorandum Bill of Lading Duplicate copy of a bill of lading.

Merchandise Car A RR car containing several less-than-carload shipments.

MESA Mining Enforcement and Safety Administration; all respirators designated for use with pesticides are jointly approved by MESA and National Institute for Occupational Safety and Health (NIOSH).

Metabolite Compound resulting from chemical action on a pesticide inside a living organism; may be more or less toxic than its "parent" pesticide; may develop in some cases when a pesticide is exposed outside a living organism.

Methane Colorless, odorless, combustible gas produced by the decomposition of vegetable matter or by chemical synthesis.

Metric System of measurement used by most of the world; meters (length), grams (weight), and liters (volume).

Metric Ton Unit of mass and weight equal to 1,000 kilograms or 2,205 pounds avoirdupois.

mg Milligram.

mg/kg Milligrams per kilogram; used to express the amount of pesticide in mg/kg of animal body weight which produces a known effect.

Micro- Prefix meaning 1/1,000,000.

Microbial Pesticide Chemical whose active ingredient is a bacteria, virus, or other tiny plants or animals.

Microgram Metric weight measurement equal to 1/1,000,000th of a gram; approximately 28,500,000 micrograms equal one ounce.

Microorganisms Plant or animal life not visible to the human eye without the aid of a microscope; found in air, water, and soil and generally include the bacteria, yeasts and fungi.

Microwave Plasma The gas generated during the detoxification reaction in an experimental hazardous waste chemical treatment process by which new stable compounds are synthesized or molecules are decomposed by microwave reactions with gas molecules.

Microwaves Radio-frequency waves generated by electronic devices in which electrons are accelerated and directed toward a target.

Midnight Dumper Person or company disposing of hazardous waste in an illegal manner.

Mileage Allowance Allowance based on distance made by railroads to owners of private freight cars.

Mileage Rate Rates applicable to distance.

Milli- Prefix meaning 1,000.

Milligram Weight measurement equal to 1/1,000 gram; approximately 28,500 mg equal one ounce.

Milling in Transit The stopping of grain, lumber, etc., at a point located between the points of origin and destination for the purpose of milling.

Milliroentgen Measure of radiation, 1/1000 of a roentgen.

Mine An excavation made in the earth to extract ores.

Mineral-Based Sorbent An inorganic substance with adsorptive or absorptive capacities used to recover oil; includes vermiculite, perlite or volcanic ash; recover 4–8 times their weight in oil.

Mineral Spirits Flammable petroleum distillates that boil at temperatures lower than kerosene; used as solvents and thinners especially in paints and varnishes; naphthas; used extensively in chemical dispersants before 1970, not in use now due to their toxicity.

Minimum Charge Least charge for which a shipment will be handled.

Minimum Rate Lowest rate that may be charged.

Minumum Weight Least weight at which a shipment is handled at a carload rate.

Mining Overburden Material overlying an economic mineral deposit which is removed to gain access to that deposit.

Ministerial A governmental decision requiring no personal judgment by a public official as to manner of execution of the project; involves only the use of fixed standards or objective measurements without personal or subjective judgment (such as automobile registrations, dog licenses and marriage licenses.)

Miscible Flood An oil recovery process involving injection of a solvent followed by a displacing fluid.

Miscible Liquids Two or more liquids that can be mixed and will remain mixed under normal conditions.

Mist Droplets of insecticide dispersed in liquid; more fine than a spray and more coarse than an aerosol.

Mist Blower Spray equipment utilizing hydraulic atomization of the liquid at the nozzle aided by an air blast past the source of spray.

Mist Drilling An oil drilling technique using air or gas to which a foaming agent has been added as a circulation medium.

Mites (Spiders, Ticks) Tiny animals related to insects; have eight jointed legs, two body regions, no antennae and no wings; mite has six legs during nymphal stage and are often grouped with ticks and spiders.

Miticide Pesticide used to control mites and ticks.

Mitigation To cause to become less harsh, to alleviate. In negotiations includes: (1) Avoiding the impact altogether by not taking a certain action. (2) Minimizing impacts by limiting the degree or magnitude of the action and its implementation. (3) Rectifying the impact by repairing, rehabilitating, or restoring the impacted environment. (4) Reducing or eliminating the impact over time by preservation and maintenance operations during the life of the action. (5) Compensating for the impact by replacing or providing substitute resources or environments.

Mixed Carload (RR) Carload of different articles in a single consignment.

Mixed Carload Rate Freight charge applicable to a carload of different articles in a single consignment.

Mixed Paper Waste paper of various kinds and quality; usually collected from stores, offices and schools.

Mixtures Substances containing two or more materials not chemically united.

MLD Median lethal dose (LD50).

Mobile Intensive Care Nurse (MICN) Authorized Registered Nurse.

Mobile Intensive Care Unit An emergency vehicle staffed by EMTs, mobile intensive care nurses and/or physicians and equipped to provide intensive care to the ill or injured at the scene of medical emergencies and during transport to a hospital.

Mode of Action Manner in which an herbicide controls weeds or plant growth; also the ways in which pesticides affect people and other mammals.

Moderator Material such as ordinary water, heavy water, or graphite, used in a reactor to slow down high velocity neutrons, thus increasing the likelihood of further fission.

Moisture Content Amount of water absorbed in a fuel at a given time; has an affect of regulating ignition time and rate of combustion of a fuel. A fuel generally absorbs moisture from the air but can release moisture to drier air, an important factor in fire behavior.

Mold A fungus-caused growth often found in damp or decaying areas or on living things.

Molded Pulp Products Contoured fiber products molded from paper pulp for protective packaging, such as egg cartons.

Molecular Weight Total mass of any group of atoms bound together to act as a single unit.

Molecule Smallest part of a substance which retains the chemical and physical properties of that substance; a combination of two or more atoms of the same or different elements.

Molluscicide Pesticide used to control slugs and snails.

Mollusks Any of a large family of invertebrate animals, including snails and slugs.

Monitoring (1) Periodic or continuous determination of radiation levels. (2) Ongoing surveillance of a hazardous materials waste disposal site by measurements or observations of the ambient air, ground water, surface water, and soil conditions.

Monitoring, Radiological Operation of locating and measuring radioactive contamination by means of survey instruments that can detect and measure (as dose rates) ionizing radiations.

Monitoring System Regular measurements of any pesticide escaping into the environment.

Monofills Landfills, surface impoundments, or waste piles used to treat, store or dispose of one or more of a small group of inorganic wastes; includes wastes that are hazardous solely because they exhibit EP Toxicity.

Moribund The condition exhibited by an organism just prior to death or recovery, especially when poisoned. Organisms in this state manifest abnormal reactions to stimuli.

Movable Fuel Storage Tenders (Farm carts) Containers in excess of 1200 gallons water capacity equipped with wheels to be towed from one location to another; basically non-highway vehicles used as fuel supply for farm tractors, construction machinery and similar equipment.

MPH Miles per hour.

Mud The liquid circulated through the wellbore during rotary oil well drilling and workover operations. Used to cool and lubricate the bit and drill stem, protecting against blowouts by holding back subsurface pressures.

Mud Acid A mixture of hydrofluoric acids and surfactants used to effect mud removal from the wellbore.

Mud Additive Any material added to drilling fluid to change some of its characteristics or properties.

Mud Analysis Testing of drilling mud to determine its chemical properties.

Mudflap Deflecting shield at rear of wheels to minimize road spray or other road surface substances from being thrown rearward; mudflaps designed to remain rigid when vehicle is traveling at high speed are termed "antisail."

Mudhop (RR slang) Yard clerk.

Mud Hose A reinforced, flexible rubber tube through which oil drilling mud flows from the standpipe.

Multiple Casualty Incident (EMS) An emergency involving two or more casualties requiring response of more than one medical transport unit with transport priority given to the most seriously injured.

Multipurpose Capable of more than one use; multipurpose pesticides may control more than one pest.

Municipal Solid Waste Non-hazardous, non-agricultural solid waste generated by residences, businesses and institutions.

Municipality A city, borough, incorporated town, township, or county, or any authority created by one of the foregoing.

Mutagen Substance causing genes in an organism to mutate or change.

Mutagenic Capable of producing a genetic change.

Mutagenic Agent A chemical agent bringing changes in the hereditary makeup of the individual when applied to a living organism, resulting in progeny differing from the parent in some respect.

Mutagenesis Alteration of the inherited genetic material, i.e., alteration of DNA in the paternal or maternal reproductive cell; may cause offspring to be born malformed.

Mutual Aid Region A subdivision of the state's emergency organization established to facilitate the coordination of mutual aid and other emergency operations within a multicounty geographical area of the state.

mV Millivolt.

Mwe Megawatts of electricity; rating of power output of a generating station.

Mycoplasma-Like Organisms Organisms recently discovered to be the cause of many plant diseases formerly attributed to viruses; organisms smaller than bacteria and larger than viruses.

Mylar A thin plastic material used in alpha survey equipment.

N

Nano- Prefix that divides a basic unit by one billion; 1/1,000,000,000.

Naptha Various volatile and often flammable liquid hydrocarbon mixtures used as solvents and diluents; consists mainly of hydrocarbons with higher boiling point than gasolines and lower boiling point than kerosene; principal component of chemical dispersants used prior to 1970.

Napthenes Class of hydrocarbons with similar physical and chemical properties to alkanes; insoluble in water, generally boil at 10–20°C higher than corresponding carbon number alkanes.

Narrow Gauge (RR Track) When the distance between the heads of the rails is less than 4 feet 8 inches.

Narrowleaf Plants, Species, or Weeds Plants having narrow leaves and parallel veins, as compared to broadleaf plants. Examples: grasses, sedges, rushes, and onions.

National Emergency Equipment Locator System (NEELS) A computerized inventory containing description, location, and who to contact information for emergency spill cleanup equipment in parts of the United States and Canada.

National Pollutant Discharge Elimination System (NPDES) A program established under the Federal Water Pollution Control Act requiring all point source discharges into any body of water to be permitted by EPA or the designated state agency; minimum pretreatment requirements for such discharges are established under this program.

Natural Circulation Reactor Nuclear reactor in which the coolant (usually water) is made to circulate without pumping, by natural convection resulting from the different densities of its cold and reactor-heated portions.

Natural Enemies Predators and parasites in the environment which attack pest species.

Natural Gas A mixture of hydrocarbons and varying quantities of nonhydrocarbons existing in natural underground reservoirs, either in a gaseous phase or in solution with crude oil.

Natural Gas Liquids Those portions of reservoir gas liquefied at the surface in leases separators, field facilities or gas processing plants.

Natural Hazards Geologic, meterological or biological conditions which affect the safety of facility operations, thereby posing potential risks to human health and the environment.

Natural Organic Sorbent Natural materials such as peat moss, straw and sawdust which can be used to recover spilled oil; generally absorb 3–6 times their weight in oil because of the criss-cross arrangement of fibers within the material. All natural sorbents will absorb water as well as oil and virtually all sink when saturated with water.

Natural Resources Land, fish, wildlife, biota, air, ground water, drinking water supplies, and other such resources belonging to, managed by, held in trust by, appertaining to or otherwise controlled by the United States, state or local government, foreign government, or private concern or individual.

Natural Uranium (Normal uranium) Uranium as found in nature, containing 0.7% 235U, 99.3% 238U, and a trace of 234U.

NCRP National Council on Radiation Protection and Measurements, an independent organization chartered by the U.S. Congress.

Necrosis Death in a particular part of a living tissue; example: death of a certain area of a leaf.

Necrotic Showing varying degrees of dead spots or areas.

Negative Declaration A written statement by a lead agency briefly describing reasons that a proposed project, not exempt from an environmental quality agency, will not have a significant effect on the environment and therefore does not require preparation of an Environmental Impact Report.

Negligence Failure to do a job or duty; an act or state of neglectfulness.

Negligible Residue A tolerance set on a food or feed crop containing a very small amount of pesticide at harvest as a result of indirect contact with the chemical.

Negotiation A process wherein parties compromise to reach a mutual agreement.

Nematicide Pesticide used to control nematodes; often applied as a soil fumigant.

Nematode Worm-like organism feeding on or in plants and animals; some are microscopic, some larger; many are internal parasites of people and other animals; common names: roundworms, threadworms and eelworms.

Neoprene Synthetic rubber used to make gloves and boots offering protection against most pesticides.

Nervous System All nerve cells and tissues in animals, including the brain, spinal cord, ganglia, nerves, and nerve centers.

Nested Packed one within another.

Net Ton 2,000 pounds.

Net Ton-Mile (RR) Movement of a ton of freight one mile.

Net Weight (1) Weight of an article clear of packing and container. (2) As applied to a carload, the weight of the entire contents of the car.

Net, Network (EMS) An orderly arrangement of stations interconnected through communications channels to form a coordinated entity.

Neutralize To destroy the effectiveness of (something); to make harmless anything contaminated with a chemical agent.

Neutralization Process in which acid or alkaline properties of a solution are altered by the addition of certain reagents to bring the hydrogen and hydroxide concentrations to an equal value; sometimes referred to as 7 pH, the value of pure water.

Neutralization Surface Impoundments Surface impoundments that: (1) contain no other wastes; (2) are used to neutralize wastes that are hazardous solely because they exhibit the characteristic of corrosivity; (3) neutralize the corrosive wastes sufficiently rapidly so there is no potential for migration of hazardous waste from the impoundment.

Neutron A component of a nucleus, 2,000 times more massive than an electron; differs from a proton by its neutral (zero) electrical charge; also a type of ionizing radiation.

Newsprint Type of paper generally used for printing newspapers.

NFPA National Fire Protection Association.

NFPA 704M A pamphlet describing a system for identifying fire hazards published by the NFPA.

NHTSA National Highway Traffic Safety Administration of Department of Transportation, responsible for establishing motor vehicle safety standards and regulations for new vehicles; formerly the National Highway Safety Bureau (NHSB).

Nine One One (911) A three digit emergency telephone number promulgated as the nationwide emergency access number.

NIOSH National Institute for Occupational Safety and Health; *see* MESA; all respirators designated for use with pesticides are jointly approved by NIOSH and MESA.

Nipple A tubular pipe fitting threaded on both ends and less than 12 in. long.

Nitrophenols Synthetic organic pesticides containing carbon, hydrogen, nitrogen, and oxygen; used as wood preservatives, fungicides, or disinfectants; affect liver and central nervous system in the human body.

NLPGA National LP-Gas Association.

NOAA National Oceanic and Atmospheric Administration.

NOAA Weather Station A mobile weather data collection and forecasting facility (including personnel) provided by the NOAA which can be utilized at a hazardous materials incident.

NOI Not Otherwise Indexed; legal classification of radioactive material transported in sole-use vehicles.

Non-Accumulative Does not build up or store in an organism or in the environment.

Non-Agency Station A railroad station which does not have an agent; a closed station.

Nonane A paraffin hydrocarbon that is liquid at atmospheric conditions; boiling point is 303.5°F.

Non-Destructive Testing Testing to detect internal and concealed defects in materials using techniques that do not damage or destroy the items being tested; X-rays, isotropic radiation and ultrasonics are frequently used.

Non-Ferrous Metals Metals which contain no iron, such as aluminum, copper, brass, bronze.

Nonflammable Gas A compressed gas not classified as flammable.

Nonliquefied Gas A gas that is entirely gaseous at a temperature of 70°F (21°C).

Non-Persistent Pesticide Pesticide that breaks down almost immediately or only lasts for a few weeks or less and turns into non-toxic by-products; may be broken down by light, moisture or microorganisms, or may evaporate.

Non-Point Source A source from which pollutants emanate in an unconfined and unchannelled manner, including but not limited to (1) water effluent not controlled through NPDES permits or traceable to a discrete identifiable origin but resulting from natural processes, such as nonchannelled runoff, precipitation, drainage or seepage; (2) air contaminant emissions from landfills and surface impoundments.

Non-Selective Pesticide A pesticide that is toxic to a wide range of pests or toxic to more than one animal or plant. Same as BROAD SPECTRUM.

Non-Target Any plant, animal, or other organism which is not the object of a pesticide application; any species that the pesticide is not aimed at.

Non-Toxic Not poisonous.

Non-Volatile A substance that does not evaporate at normal temperatures when exposed to the air.

Normal Speed Maximum authorized speed shown in the timetable.

NOS Not Otherwise Specified, a legal classification of radioactive material transported in sole-use vehicles.

Noxious Weed A plant considered to be especially unwanted or troublesome.

Nozzles Devices which control drop size, rate, uniformity of coverage; in use with a pesticide, they also determine the thoroughness, and safety of a pesticide application. Nozzle type determines the ground pattern of coverage.

Nuclear An adjective referring to the atom's nucleus.

Nuclear Battery A radioisotopic generator.

Nuclear Chain Reaction A sequence of fissions in which neutrons, released during the spontaneous disintegration of a nucleus, collide with other nuclei and cause these in turn to disintegrate and release more neutrons.

Nuclear Energy Energy liberated by a nuclear reaction (fission or fusion) or by radioactive decay.

Nuclear Explosive An explosive based on fission or fusion of atomic nuclei.

Nuclear Power Plant Any device, machine, or assembly that converts nuclear energy into some form of useful power, such as mechanical or electrical; in a nuclear electric power plant, heat produced by a reactor is generally used to make steam to drive a turbine that in turn drives an electric generator.

Nuclear Reaction Change in an atomic nucleus, such as fission, fusion, neutron capture, or radioactive decay, as distinct from a chemical reaction, which is limited to changes in the electron structure surrounding the nucleus.

Nuclear Reactor A device in which a fission chain reaction can be initiated, maintained, and controlled; essential component is a core with fissionable fuel; usually has a moderator, a reflector, shielding, coolant, and control mechanisms.

Nuclear Transformation The change in an isotope from a less stable to a more stable form, typically involving the emission of alpha, beta, gamma, and/or neutron radiation.

Nuclear Weapon A device capable of producing a nuclear yield by either fission, fusion, or both; contains conventional explosives.

Nucleonics The science and technology of nuclear energy and its applications.

Nucleus The center part of an atom, composed of protons and neutrons. Plural: nuclei.

Nuclide A general term applicable to all atomic forms of the elements; term is often erroneously used as a synonym for "isotope" which properly has a more limited definition. Whereas isotopes are the various forms of a single element (so are a family of nuclides) and all have the same atomic number and number of protons, nuclides comprise all the isotopic forms of all the elements. Nuclides are distinguished by their atomic number, atomic mass, and energy state.

Nymph The stage of development in certain insects after hatching when the young insect looks like an adult but lacks fully developed wings.

O

Objective The main purpose to be achieved by tactical units at a hazardous materials emergency.

Obsolete Scrap Scrap materials derived from products which have completed their useful economic life.

Octane Rating Classification of gasoline by its antiknock qualities; the higher the octane number or rating, the greater the antiknock qualities of the gasoline.

Occupational Safety and Health Administration (OSHA) An agency of the United States government charged with establishing and enforcing safety standards for industry employees.

OD Outside diameter.

Odorant A chemical added to natural gas to enable the presence of gas to be detected by smell.

Office Car Car used by railway officials while traveling.

Office of Emergency Services (OES) A California state agency responsible for assisting local jurisdictions in preparing for, responding to, and recovering from the effects of a disaster.

Off-Scene Support Assistance (via telephone, radio, or computer) from technical persons, agencies, shippers, responders, etc., not at the accident site.

Offshore Drilling Drilling for oil in an ocean or large lake; drilling unit may be a mobile floating vessel with a ship or barge hull, a semisubmersible or submersible base, a self-propelled or towed structure or a permanent structure.

Off-Site Facility A recycling, treatment, or disposal facility located at a site other than where the wastes are generated; often operated by a firm whose business is solely for the purpose of processing industrial wastes and which itself generates little or no hazardous waste.

Off-Site Hazardous Waste Facility An operation involving handling treatment, storage, or disposal of hazardous wastes such that the waste is transported commercially to the site.

Oil and Gas Separator Production equipment used to separate liquid components of the oil well stream from the gaseous elements. Separators are vertical or horizontal and cylindrical or spherical in shape. Separation is accomplished principally by gravity, the gas rising to the top and heavier liquids falling.

Oil Hazardous Materials Technical Assistance Data System Organization within the EPA that provides information on some hazardous substances to emergency terms responding to spills.

Oil in Water Emulsion An emulsion of oil droplets dispersed in surrounding water, formed as a result of wave action or by a chemical dispersant; show a tendency to coalesce and reform an oil slick when the water becomes calm.

Oils Liquids used to carry an active ingredient in a pesticide formulation or to weaken a pesticide formulation, or for direct application as a pesticide.

Oil Saver A gland arrangement operated either hydraulically or mechanically, used to prevent leakage by pressure sealing around wire rope when swabbing an oil well.

Oil Seep A surface location where oil appears having permeated from its subsurface boundaries and accumulated in small pools.

Oil Slick Film of oil floating on water, considered a pollutant.

Oil Slug Name given to a downward-moving oil mass which often results when oil is spilled on relatively porous soil; the slug-like shape results from the tendency of the mass to leave behind a funnel of soil which is partially saturated with oil.

Oil Spill Contractor Private firms which have been formed to provide oil spill cleanup services; often have their own containment and recovery equipment and may be identified in local and regional contingency plans.

Oil Spill Cooperative Organizations formed by oil companies operating in a given area for the purpose of pooling equipment and training personnel to combat oil spills.

Oleophilic Agent A material or chemical which has the tendency to attract oil; chemicals used to treat sorbent materials to increase their oil recovery capacity.

On the Ground On the ties, not on the rails, as a derailed train.

On Scene Commander (OSC) The overall coordinator of an oil spill response team, usually a representative of an oil company, a government official, or an independent oil spill cleanup contractor; responsible for on-site strategical decisions and actions throughout each phase of a cleanup operation and who maintains close liaison with the appropriate government agencies to obtain support and provide progress reports on each phase of the emergency response.

On-Site Facility A treatment process or facility located on or adjacent to the premises of a waste generating firm and usually operated by that firm.

Open Burning The burning of waste materials in the open or in a dump; produces smoke, odor and other objectionable air pollutants.

Open Cycle Reactor System A reactor system in which the coolant passes through the reactor core only once and is then discarded.

Open Dump An open land site where solid waste is handled and disposed of improperly; usually the site is exposed to the elements, harbors insects and rodents, and leaks methane gas. The Federal Resource Conservation and Recovery Act (RCRA) authorizes the State Solid Waste Management Board to inventory open dumps in California as a step in enforcing minimum waste disposal standards.

Open Dump Inventory A survey undertaken by the State Solid Waste Management Board's enforcement division, in cooperation with the Federal Environmental Protection Agency, to determine the number, location and regulation compliancy of all disposal facilities and open dumps in California; this inventory will help the Board put an end to improper solid waste disposal practices.

Operator The person responsible for the overall operation of a facility.

Operating Speed The constant rate at which a pesticide sprayer moves during application; usually measured in miles per hour or feet per minute.

Operational Period (ICS) A period of time scheduled for execution of a given set of suppression and rescue actions as specified in the Incident Action Plan.

Operational Radiation Monitoring Systems (ORMS) An instrumentation system at a nuclear reactor designed to detect and alarm abnormal radiation levels in process and effluent streams.

Opposing Signals Signals which govern movements in opposite directions on the same railroad track.

Oral Of, by, through, or into the mouth.

Oral Toxicity How poisonous a pesticide is to an animal or person when taken by mouth.

Orange Oxide Uranium trioxide.

Ordinate The vertical plane on graphs usually expressed in increasing order of magnitude from bottom to top. In pesticides, consisting of probability units or percent mortality of test organisms and plotted against the observed mortality.

Organic Chemical compound containing carbon; may occur in nature or be produced by chemical synthesis.

Organic Chemistry The study of the compounds of carbon.

Organic Cooled Reactor A reactor that uses organic chemicals such as mixtures of polyphenyls (diphenyls and terphenyls) as coolant.

Organic Matter Chemical substances containing the element carbon, originating in animal or plant life or their derivatives, coal, or petroleum.

Organic Peroxide An organic derivative of the inorganic compound hydrogen peroxide.

Organic Pesticides Pesticides which contain carbon; major groups are petroleum oils and synthetic organic pesticides.

Organic Phosphate Insecticide Examples include: Parathion, HETP and TEPP.

Organism Any living thing.

Organochlorine Compounds Synthetic organic pesticides that contain chlorine, carbon and hydrogen; affect the central nervous system.

Organophosphates Synthetic organic pesticides that contain carbon, hydrogen, and phosphorous; highly toxic to humans as they prevent proper transmission of nerve impulses.

Organs Body parts affected by substance exposure.

Original Container The package (bag, can, bottle, etc.) prepared by the manufacturer in which the pesticide is placed and then sold; package must be labeled with what the pesticide is, how to use it safely and correctly, and how to legally dispose of the empty container.

ORMS Other Regulated Materials; materials that do not meet the definitions of hazardous materials, but possess enough hazard characteristics that they require some regulation; relating to ORM-A, -B, and -C; ORM-D materials are hazardous materials transported in small quantities.

Osmosis The tendency of a fluid to pass through a semipermeable membrane typically separating a solvent and a solution so as to tend to equalize their concentrations on both sides of the membrane.

Outcome Result.

Outer Marker Beacon An aircraft radio beacon defining the first predetermined point during an instrument landing system approach.

Outlet Valve The valve farthest downstream in a tank piping system to which the discharge hose is attached.

Out of Service Resources (ICS) Material or personnel resources assigned to an incident but unable to respond for mechanical, rest or personnel reasons.

Outrigger Structural load-carrying members attached to and extending outward from the main longitudinal frame members of a trailer.

Overflow Pipe Device installed at the top of a tank to enable the liquid within to be discharged to another vessel when tank is filled to capacity.

Overhead Personnel Supervisory personnel.

Overrun The hard surface extending past the designated end of a runway.

Over the Top Pesticide application over the top of a growing plant.

Overpressure (1) The transient pressure over and above atmospheric pressure caused by a shock wave from a nuclear explosion. (2) Pressure in excess of the design or operating pressure.

Overstress To stress an object or person beyond recoverable limits.

Overturn Protection for fittings on top of a tank in case of roll-over; may be combined with flashing rail or flashing box.

Ovicide Pesticide used to control eggs, particularly insect, mite, or nematode eggs.

Owner The person or municipality who is the owner of record of a facility or part of a facility.

Oxidation (1) The combining of oxygen with another substance chemically; when process is slow, no apparent heat or light is produced as in rusting of metal or yellowing of paper; when rapid, oxidation is accompanied by the production of heat and light (as in fire). (2) In hazardous waste management, a process to treat a waste stream with a strong oxidizing agent changing the waste chemically to a less hazardous state.

Oxidizer Substance that yields oxygen readily to stimulate the combustion of organic matter and inorganic matter.

Oxidizing Ability The ability to yield oxygen readily to stimulate combustion.

Oxidizing Agents Substance containing chemically available oxygen.

Oxygen An element which readily unites with materials.

P & A Plugged and abandoned (oil well).

Package Car RR car containing several less-than-carload shipments.

Package Freight Merchandise shipped in less-than-carload quantities.

Package Markings The descriptive name, instructions, cautions, weight or specification marks required to be placed upon the outside of containers of hazardous materials.

Package Power Reactor A small nuclear power plant designed to be crated in packages small enough to be conveniently transported to remote locations.

Packaging A broad term used by the Department of Transportation to describe shipping containers, and any markings, labels, or placards affixed to them.

Packaging Materials Any of a variety of papers, cardboards, metals, wood, paperboard and plastics used in the manufacture of containers for food, household, or industrial products.

Pair Production The transformation of the kinetic energy of a high-energy photon or particle into mass, producing a particle and its antiparticle, such as an electron and a positron.

Pallet A small portable platform for holding material for storage or transportation.

Pantograph A device located on top of electric RR equipment which collects power from an overhead contact wire by means of a sliding contact shoe.

Paper A thin sheet material made of cellulose pulp, derived mainly from wood, rags, and certain grasses, processed into flexible leaves or rolls by deposit from an aqueous suspension, used chiefly for writing, printing, wrapping and sanitary purposes.

Paperboard Heavier in weight, thicker, and more rigid than paper. Basic classes of paperboard are: (1) container board, (2) box board and (3) special types such as automobile board, building board, etc.

Paper Rate A published rate under which no RR traffic moves.

Paperstock General term used to designate waste papers which have been sorted or segregated at the source into various grades.

Paraffin A waxy substance obtained from the distillation of crude oils; complex mixture of higher carbon number alkanes that is resistant to water and water vapor and is chemically inert.

Paraffin Inhibitor A chemical injected into an oil well bore to prevent or minimize paraffin deposition.

Parasite An organism that lives and feeds on or in another plant or animal (known as the host) which is usually harmed by the parasite. Parasites are usually pests, but when used to attack other pests which can injure crops or animals, they become forms of biological control.

Parasitic Living off of or feeding on or in another organism.

Partial Body Exposure The exposure to radiation of only an isolated part of the body such as an arm, foot, or head, rather than the whole body.

Partial Closure Securing a portion of a facility when it is filled to capacity to prevent environmental damage while the remainder of the site is being filled.

Participating Carrier (Tariff) A railroad which is a party, under concurrence, to a tariff issued by another railroad or by a tariff publishing agent.

Particle A minute constituent of matter, generally one with a measurable mass.

Particulate Matter Normally refers to dust and fumes; travels easily through air.

Parts Per Million Permissible pesticide tolerance on crops is expressed in parts per million (ppm); one ppm is equal to one inch in 16 miles.

Pathogen Any disease-producing organism.

Pawl Ratchet; on RR brake wheels, a pivoted bar adapted to fall into the notches or teeth of a wheel as it rotates in one direction and to restrain it from backward motion.

Pebble Beach A beach substrate composed primarily of gravel; more coarse than sand and can allow stranded oil to penetrate to a considerable depth.

Penetrant An adjuvant which helps a liquid pesticide get into the pores of or get through the outer surface of a leaf or root and into the plant; also called penetrating agent.

Penetration The ability to get through; the process of entering.

Peptization A method of getting substances into colloidal suspension by breaking down larger particles.

Percent by Weight A percentage expressing the active ingredient weight as a part of the total weight of the formulation.

Percent Concentration The weight or volume of an active ingredient expressed as a percentage in the final formulation.

Percolation The movement, flow or infiltration of water through the pores or spaces of rock or soil.

Per Diem A charge made by one transportation line against another for the use of its cars; based on a fixed rate per day.

Perforating Gun Device fitted with shaped charges, then lowered to the desired depth in an oil well and fired to create penetrating holes in casing, cement, and formation.

Periodic Table (Periodic Chart) A list of all elements arranged in order of increasing atomic numbers and grouped by similar physical and chemical characteristics into "periods"; based on the chemical law that physical or chemical properties of the elements are periodic functions of their atomic weights.

Perishable Commodities easily spoiled or damaged because of weather or delay in transit; usually describing foodstuffs.

Permeability The property of soil or rock allowing passage of water through it; depends not only on the volume of openings and pores, but also on how these openings are connected to each other.

Permeable Open to passage or penetration; used especially for a substance that allows the passage of fluids.

Permissible Exposure (Maximum) That dose of ionizing radiation established by competent authorities as an amount below which there is no reasonable expectation of risk to human health, and which at the same time is somewhat below the lowest level at which a definite hazard is believed to exist.

Permits Official approval and permission to proceed with an activity controlled by the permitting authority.

Persist To remain or stay.

Persistent Term applied to a pesticide that remains in the environment for a fairly long time.

Personal Alarm Dosimeter A type of dosimeter which emits an audible alarm whenever a preset threshold value of radiation is reached.

Personal Protective Clothing Those items of attire as recommended for protection against hazardous materials, including protective coats, trousers, boots, face and head masks, and gloves.

Personnel Decontamination Removal of radioactive contamination from individuals; removal of hazardous substance from individuals.

Personnel Shield A recognizable structure on transportation vehicles which keeps people from accidentally getting too close to the radioactive cargo, thus protecting them against exposure.

Perturbation An event or unexpected encountered condition capable of throwing an activity into disorder.

Pest An unwanted organism; a living thing that competes with people for food and fiber or attacks people directly.

Pesticide A chemical or mixture of chemicals used to destroy, prevent, or control any living organism considered to be a pest; any substance used to control or destroy insects, weeds, rodents, disease-causing organisms, and other types of pests which attack living things or spread disease among plants and animals.

Pesticide Chemical A pesticide which is a chemical rather than a non-chemical pesticide (like a parasite or virus).

Pesticide Kill The death of a large number of non-target organisms due to careless or improper application or use of a pesticide.

Pesticide Tolerance Amount of pesticide residue which may legally remain on or in a food crop, as established by the EPA; *see* RESIDUE.

Petroleum Oils Pesticides used to control insects and plants refined from crude oil.

Petroleum Products Pesticides and other substances containing oil, gasoline, kerosene, or similar products.

pH A numerical designation of relative acidity and alkalinity; a pH of 7.0 indicates precise neutrality, high values indicate increasing alkalinity and lower values indicate increasing acidity.

Phase I RCRA Regulations promulgated in May 1980 including the identification and listing of hazardous wastes, standards for generators and transporters of hazardous wastes, standards for owners and operators of facilities that treat, store, or dispose of hazardous waste; requirements for obtaining hazardous waste facility permits, and rules governing delegation of authority to the states.

Phase II RCRA Technical requirements for permitting hazardous waste facilities; sets specific standards for particular types of facilities to ensure the safe treatment, storage and disposal of hazardous waste on a permanent basis by methods that will protect human health and the environment. Phase II standards enable facilities to move from interim status to final facility permits.

Pheromones Chemicals produced by insects and other animals to communicate with other members of the same species; some are used to monitor insect populations; most pheromones used today are synthetic.

Photodegradable Able to decompose through a chemical reaction initiated by direct exposure to the sun's ultraviolet light.

Photon A minute, massless packet of pure light energy; all electromagnetic radiation, including gamma radiation, is in the form of photons.

Physical Description A brief summarization of the form of a substance, specifying whether it is solid, powder, flakes, crystals, liquid, gas, etc., accompanied by identification of color and odor where applicable.

Physical Properties Properties of a material relating to the physical states common to all substances, i.e., solid, liquid, or gas.

Physical Treatment The process by which waste is rendered nonhazardous by physically removing the hazardous substance from the waste stream, or is rendered more readily disposable or transportable by reducing the water content or solidifying the waste. Current methods of physical treatment are: (1) Sedimentation removing suspended solid particles by providing time and space for settling in special tanks or holding ponds. (2) Filtration separating liquids and solids by using various types of filters. (3) Solar

Evaporation reducing liquid wastes in volume through simple evaporation in uncovered impoundment ponds. (4) Distillation separating liquids with different boiling points by heating the mixture to vaporize and retrieve certain components.

Phytotoxic Poisonous to plants.

Pick-Up Plate A sloped structure located forward of the kingpin of a trailer, designed to facilitate engagement of the fifth wheel to the kingpin.

Pico- Prefix meaning one-trillionth; 1/1,000,000,000,000.

Pig A Type B container; usually lead, used to ship or store radioactive materials; thick walls provide protection for handlers.

Piggy-Back A type of shipping in which bulk containers from one mode, such as highway transportation, are placed on flat cars or container ships for transportation by another mode, such as rail or marine.

Pile (1) A non-containerized accumulation of solid, non-flowing hazardous waste. (2) Old term for NUCLEAR REACTOR, as first reactor was built by piling up graphite blocks and natural uranium.

Pilot (RR) Employee assigned to a train when the enginemen or driver of a track car is not qualified on the physical characteristics or rules of the railroad.

Pin Puller (slang) A trainman who uncouples cars while switching by lifting the coupler pin with the uncoupling lever located on each end of a car.

Pintle Hitch.

Pipeline A product discharge line.

Pipeline Oil Oil clean enough to be acceptable to transport or purchase; a crude oil acceptable for pipeline shipment.

Piping, Piping Systems Pipe, tubing, hose and flexible rubber or metallic hose connectors made up with valves and fittings into complete systems for conveying substances in either liquid or vaporized state at various pressures from one point to another.

Pisicide Pesticide used to control fish.

Piston Travel (RR Air Brake) The amount of piston movement when forced outward as the brakes are applied.

Pitot Tube (1) Device with an open end facing upstream, used to measure the flow of air past an aircraft. (2) In the petroleum industry, used to measure the velocity head of a flowing medium.

Pivot Pin Coupler.

Placards (1) Diamond-shaped markers 10-3/4″ square required on a transporting vehicle such as a truck or tank car or a freight container 640 cu ft or larger. (2) Diamond-shaped sign required on outside of vehicles transporting radioactive materials displaying same standard warning terms and symbols as a label. (3) Paper forms of various designs used to identify RR cars requiring special attention (dangerous, explosives, etc.).

Plane Source Layer of radioactive sources spread over a given area.

Planning Meeting (ICS) A consultation held as needed throughout the duration of an incident to select specific strageties and tactics for incident control operations and for service and support planning.

Planning Section Chief (ICS) Individual responsible for comprehension of current situation, predicting probable course of the incident, preparing primary and alternate strategies for the Incident Commander, and for collecting, evaluating and disseminating information about the incident.

Plant Factor Ratio of the average power load of an electric power plant to its rated capacity; capacity factor.

Plasma An electrically neutral gaseous mixture of positive and negative ions; sometimes called the "fourth state of matter," since it behaves differently from solids, liquids and gases. High-temperature plasmas are used in controlled fusion experiments.

Plastics Man-made materials consisting of large molecules called "polymers" containing primarily carbon and hydrogen with lesser amounts of oxygen or nitrogen, frequently compounded with various organic and inorganic compounds as stabilizers, colorants, fillers and other ingredients.

Plowshare The Atomic Energy Commission program of research and development on peaceful uses of nuclear explosives; possible uses include large-scale excavation such as for canals and harbors, crushing ore bodies, and producing heavy transuranic isotopes. Term is based on a Biblical reference: Isaiah 2:4.

Plug Back To shut off lower formation in a well bore.

Plug Door Door on a refrigerated RR car or box car which is flush with side of car when closed; to open, it is swung out and rolled to one side; sliding flush door.

Plume (1) The column of non-combustible products emitted from a fire (smoke). (2) A vapor cloud formation having shape and buoyancy. (3) The airborne radioactive material released from a Nuclear Power Plant and carried by the prevailing winds which may affect radiologically those downwind areas over which it passes.

Plume Exposure Pathway Route by which the radioactive material released from a nuclear facility may expose the population-at-risk to radiation; exposure may be external from the passing plume, from contaminated surfaces, or from inhalation of the passing plume.

Plunger Lift A method of lifting (oil) using a swab or free piston propelled by compressed gas from the lower end of the tubing string to the surface.

Plutonium A heavy radioactive, synthetic metallic element with atomic number 94; used for reactor fuel and in weapons.

Point of Origin (RR) Station at which a shipment is received by the railroad or freight handler from the shipper. (2) In a fire, the starting point of ignition.

Point of Transfer (LP-Gas) Location where connections are made or where LP-Gas is vented to the atmosphere in the course of transfer operations.

Point Source (1) One piece of material which is emitting all the radiation in an area; distinguished from ISOLATED SOURCE and PLANE SOURCE. (2) A discernible, confined and discrete conveyance including but not limited to a pipe, ditch, channel, tunnel, conduit, well, discrete fissure, container, rolling stock, concentrated animal feeding operation, or vessel or other floating craft from which pollutants are or may be discharged. Does not include return flow from irrigated agriculture.

Poison A substance that through its chemical action kills, injures, or impairs an organism.

Poison Control Center An information source for human poisoning cases, including pesticides, usually located at major hospitals.

Poisonous Bait A food or other substance mixed with a pesticide so a pest will be attracted to it, will eat it, and then be killed by it.

Polarized Couplers Fittings for connecting of air brake lines between vehicles; service and emergency couplings are unilateral and will not mate with each other.

Pollinators Bees, flies, and other insects which visit flowers and carry pollen from flower to flower.

Pollutant A harmful chemical or waste material discharged into the water, soil, or atmosphere; an agent that makes something dirty or impure.

Pollute To add an unwanted material (often a pesticide) which may do harm or damage; contaminate, make unclean or unsafe for use.

Pollution Contamination of air, water, land or other natural resources that will be or is likely to create a public nuisance or to render such air, water, land or other natural resources harmful, detrimental or injurious to public health, safety or welfare, or to domestic, municipal, commercial, industrial, agricultural, recreational or other legitimate beneficial uses, or to livestock, wild animals, birds, fish or other life.

Polychlorinated Biphenyls A series of hazardous chemical compounds which have been manufactured for more than 40 years for such common purposes as electrical insulation and heating/cooling equipment. Now suspected to be carcinogens, PCBs have been disposed of in the air, on land and in water; recent surveys have detected the presence of PCBs in every part of the country, even those remote from PCB manufacturers.

Polyethylene A polymer taking the form of a lightweight thermoplastic with high resistance to chemicals, low water absorption, and good insulating properties; manufactured in a number of forms, have been used with considerable success as a sorbent for oil spill cleanup.

Polymerization Reactions in which two or more smaller molecules chemically combine to form larger molecules; reaction is often violent.

Polyurethane Any of a class of synthetic resinous, fibrous or elastomeric compounds belonging to the family of organic polymers, consisting of large molecules formed by the chemical combination of successive smaller molecules into chains or networks; best known are the flexible foams used as upholstery material and mattresses.

Polyvinyl Chloride (PVC) Common plastic material which, when burned, releases toxic hydrochloric acid.

Pool Car (RR) Specially equipped cars of different ownerships assigned to a specific company or location.

Pool Reactor A reactor in which the fuel elements are suspended in a pool of water that serves as the reflector, moderator, and coolant; used for research and training.

Population-at-Risk Those persons for whom protective actions are being planned, taken, or would be taken.

Porosity The percentage that the volume of the pore space bears to the total bulk volume; in sand or sandstone, determines the amount of space available for storage of fluids.

Port An opening for access; in nuclear reactors, an opening through which objects are inserted for irradiation or from which beams of radiation emerge for experimental use.

Portable Container (LPG) A vessel designed to be readily moved, as distinguished from containers designed for stationary installations.

Portable Storage Container (LPG) Vessel similar to, but distinct from those designed and constructed for station installation, designed so it can be readily moved over highways substantially empty of liquid from one usage location to another; has legs or other supports attached or mounted on running gear (trailer) with suitable supports permitting them to be placed or parked in a stable position on a reasonably firm surface. For large volume, limited duration product usage (such as at construction sites and normally for 12 months or less) these vessels function in lieu of permanently installed stationary containers.

Portable Tank (Skid Tank) (LPG) Container of more than 1,000 pounds water capacity used to transport LP-Gas handled as a "package," i.e., filled to its maximum permitted filling density; mounted on skids or runners, with all CONTAINER APPURTENANCES protected in such a manner that they may be safely handled as a "package."

Port of Entry Location at which foreign goods are admitted into a receiving country.

Positron A particle also called a beta particle, produced when a proton transforms to a neutron; identical to an electron except for its opposite electrical charge of +1.

Post A vertical structural member.

Post-Consumer Scrap Any uncontaminated packaging material that is recoverable such as tin cans, egg cartons, glass bottles/jars.

Post-Emergent Pesticide used to control a crop or weed after it has appeared.

Potency Chemically or medicinally effective; strength; rate of toxicity.

Potential Energy Energy derived from position rather than motion, with reference to a specified field of force.

Potential Test A measurement of the maximum rate at which something can be utilized.

Potentiate To activate.

Potentiation The increase in toxicity (usually considered an undesirable effect) of a pesticide when combined with one or more pesticides.

Pour-On A pesticide which is poured along the midline of the backs of livestock.

Pour Point The lowest temperature at which a substance, such as oil, will flow under specified conditions; important in terms of cleanup since free-flowing oils rapidly penetrate most substrates, whereas semi-solid oils tend to be deposited on the surface and will only penetrate if material is coarse or the ambient temperature high.

Power Density Rate of heat generated per unit volume of a reactor core.

Power Reactor A nuclear reactor designed to produce useful nuclear power, as distinguished from reactors used primarily for research or for producing radiation or fissionable materials.

Pozzolanic Chemical reaction between a silica-alumina-containing solid which produces a concrete-like structure from the separate particles, such as sand and lime.

ppb Parts per billion.

ppm Parts per million.

Pradicide Pesticide used to control vertebrate pests.

Precious Metals Recoverable gold, silver, or platinum.

Precipitate An insoluble solid which has been formed in a liquid by chemical action.

Precipitation A hazardous waste chemical treatment method by which dissolved material falls out of the waste solution; process is enhanced by the addition of chemicals which induce precipitation.

Prehospital Time (EMS) Interval between activation of the emergency medical transport response to an incident and arrival of the patient at a receiving facility.

Preigniton In an internal-combustion engine, a condition characterized by a knocking sound and caused by the fuel-air mixture having been ignited too soon because of an abnormal condition.

Pressure Regulator Device for maintaining pressure in a line, downstream from the valve.

Pressurized Gases Gases which are stored in pressure cylinders and remain gases at normal temperature.

Pressurized Water Reactor A power reactor in which heat is transferred from the core to a heat exchanger by water kept under high pressure to achieve high temperature without boiling in the primary system; steam is generated in a secondary circuit.

Preventative Actions Directions given by the Incident Commander at an emergency to prevent the problem from increasing.

Primacy The assumption of a state, with the approval of EPA, of the responsibility to administer and enforce a federal program, such as a hazardous waste disposal program.

Primary Assembly Area An area designated for the assembly of specific groups of people involved in an incident.

Primary Materials Virgin or new materials used for manufacturing basic products, such as wood pulp, iron ore, silica sand and bauxite.

Primary Transport (EMS) Transportation of an emergency patient from the scene of an incident to a receiving facility.

Private Applicator A person who has been certified by the Dept. of Food and Agriculture or the County Commissioner's office to apply, use or supervise the use of any pesticide which is classified for restricted use for purposes of producing any agricultural commodity on property owned or rented by the applicator or the applicator's employer or, if applied without compensation other than trading of personal services between producers of agricultural commodities, on the property of other persons.

Procedures Preplanned detailed directions for dealing with specific occurrences.

Processing Technology used to reduce volume or bulk of municipal or residual waste or to convert part or all of such waste materials to off-site reuse. Include transfer facilities, compost facilities and resource recovery facilities.

Prodromal Period The initial period of acute radiation syndrome; period in which the signs and symptoms are evident, generally from 1–4 days after exposure. In medicine, a prodrome is an early symptom which indicates the start of a disease course.

Producing Zone The formation from which oil or gas is produced in an oil field.

Product In pesticides, the substance as it is packaged and sold; usually contains an active ingredient plus adjuvants.

Production Log A well-logging method to measure and record the flow of fluid past an indicating device placed at varying depths in a producing or injection well.

Products Cycle The sequence or order in which a number of different products are batched through a pipeline.

Products Line A pipeline used to ship refined products.

Products of Combustion Chemicals, gases and substances left after a material has entered into the combustion process. Depending on what is burning those products may be smoke, tar, ash, carbon dioxide or carbon monoxide, toxic gases, and/or condensed steam.

Proficiency Test (EMS) A practical examination designed to measure the performance of skills.

Projected Dose An estimated radiation dose which the population-at-risk may potentially receive as a consequence of a nuclear incident.

Prolonged Exposure More than a brief (or one-time) contact with a hazardous material such as radioactivity or a pesticide or the residue of that material.

Promptly Available (EMS) A medically prudent period of time proportionate to the patient's condition and such that any interval of time between arrival of the patient at the facility and the arrival of the health care personnel should not be deleterious to the patient.

Propellant A liquid in self-pressurized containers that forces the active ingredient from the container.

Properties The characteristics or traits which describe a material.

Proportioning Valve A device used to balance or divide the air supply between the aeration system and the discharge manifold.

Propulsion Mechanism The process by which matter or energy is propelled from its position at rest toward another location.

Protectant A pesticide that is applied before pests are actually found but where they are expected; a preventative.

Protected Area (1) The area surrounding a hazardous materials incident which is under the control of the responsible agency and from which unauthorized persons are prohibited. (2) In nuclear site incidents, the fenced area surrounding the nuclear steam supply system, turbine-generator, and the Administration and Control Buildings. (3) An area set aside for the preservation of the ecology, such as a wild-life area.

Protective Actions Procedures taken during or after a hazardous materials incident for the protection of the general public from exposures occurring as a consequence of the incident.

Protective Action Guides Projected general public radiological dose rate or dose commitment criteria which provide guidance to local county and/or State officials for the implementation of protective actions for the protection of the general public from excessive radiation exposure following a nuclear incident.

Protective Barriers Physical boundaries used to mark the outer edge of an incident area. In radiation incidents, generally used to control access to the area, to limit exposure, and to prevent spread of contamination.

Protective Clothing Garments and accessories which prevent the contamination of underlying body surfaces or personal clothing.

Protective Equipment or Gear Any clothes, materials, or devices offering protection from contaminations, whether radioactive or toxic, or from other hazards such as fire.

Proton A component of a nucleus, 2000 times more massive than an electron; differs from a neutron by its positive (+1) electrical charge. The atomic number of an atom is equal to the number of protons in its nucleus.

Provisions Specific emergency OSHA-required equipment and procedures that the employer must provide.

psi, psia, psig Pounds per square inch, pounds per square inch absolute, and pounds per square inch gage, respectively.

PSTN Pesticide Safety Team Network. Regional teams of the National Agricultural Chemical Association designed to assist with pesticide incidents.

Public Hearing/Meeting A formal means to inform the public about impending federal, state, or local government actions, and to receive their comments.

Public Information Officer Individual on the Command Post staff responsible for providing information to the news media and others who need to know; handles inquiries about the incident.

Public Safety Agency A functional division of a public agency which provides fire fighting, police, medical or other emergency services.

Public Service Commission A state body having control of or regulating public utilities.

Publishing Agent A person authorized by transportation lines to publish tariffs of rates, rules and regulations for a railroad.

Pull the Pin (1) Uncouple a RR car (by pulling up the coupling pin). (2) An expression meaning to leave a job, resign, or retire.

Pulse An electrical signal arising from a single event of ionizing radiation.

Pulse Amplifier A device designed specifically to amplify the intermittent signals of a radiation detection instrument, incorporating appropriate pulse-shaping characteristics.

Pulsed Reactor A type of research reactor with which repeated, short, intense surges of power and radiation can be produced. The neutron flux during each surge is much higher than could be tolerated during a steady-state operation.

Pulp Fiber material produced by chemical or mechanical means from fibrous cellulose raw material and from which paper and paperboard are made.

Pump Off To pump so rapidly that the oil level drops below the standing valve on the pump.

Pump-Off Line A pipeline which usually runs from the tank discharge openings to the front of the trailer; most pumps are mounted on the tractor.

Pump Station An installation built at intervals along an oil pipeline to contain storage tanks, pumps, and other equipment to route and maintain the flow of oil.

Pump Valve Any of the valves on a reciprocating pump (as the suction and discharge valves) or on a sucker-rod pump (as a ball-and-seat valve).

Pusher (RR) An extra engine at the rear of a train used to assist a train in climbing a grade.

Pyrolysis The process of chemically decomposing an organic substance by heating in an oxygen-deficient atmosphere. High temperatures and closed chambers are used. Major products from pyrolysis of solid waste are water, carbon monoxide, and hydrogen. Some processes produce an oil-like liquid of undetermined chemical composition; gas may contain hydrocarbons and frequently there is process residue of a carbon char. All processes leave a residue of inorganic material. Gaseous products cannot be mixed with natural gas in principal distribution systems unless there is additional chemical processing. Applied to solid waste, pyrolysis has the features of effecting major volume reduction while producing storable fuels.

PWR Pressurized Water Reactor, a type of reactor used in nuclear power plants.

Pyrophoric A characteristic of those materials which, if ground into fine particles, will spontaneously ignite when exposed to air.

Pyrophoric Liquid Any liquid that ignites spontaneously in dry or moist air at or below 130°F (54°C).

Q

Q Unit used to express very large energy figures. One Q equals one billion BTUs.

Quality Factor A number that accounts for the differing ionization potentials of alpha, beta, and gamma radiation; the factor by which absorbed dose is to be multiplied to obtain a quantity that expresses on a common scale the irradiation incurred by exposed persons.

Quantum Unit quantity of energy according to the Quantum Theory. The photon carries a quantum of electromagnetic energy.

Quantum Theory The statement by German physicist Max Planck that energy is not emitted or absorbed continuously but in units or quanta; corollary is that the energy of radiation is directly proportional to its frequency.

Quench To limit or stop the electrical discharge in an ionization detector.

Quicklime Calcium oxide.

R

R Rankine temperature scale.

Rabbit (1) A device to move a sample rapidly from one place (such as inside a research reactor) to another place (such as a radiochemistry laboratory); often are small cylinders of aluminum or plastic, moved by air pressure through a long pipe. (2) Small plug run through an oil flow line by pressure to clean the line or test for obstructions.

RCRA Federal Resource Conservation and Recovery Act.

RAD Radiation Absorbed Dose; basic unit of absorbed dose of ionizing radiation; the absorption of 100 ergs of radiation energy per gram of absorbing material.

Radiation The emitting of energy from an atom in the form of particles or electromagnetic waves; energy waves that travel with the speed of light, and upon arrival at a surface are either absorbed, reflected or transmitted; not absorbed by air and travel in straight lines; are the most serious causes of fire spreading from one structure to another. Nuclear radiation is that emitted from atomic nuclei in various nuclear reactions.

Radiation Accidents Accidents resulting in the spread of radioactive material or in the exposure of individuals to radiation.

Radiation Area Any accessible area in which the level of radiation is such that a major portion of an individual's body could receive in one hour a dose in excess of 5 millirem or in any 5 consecutive days a dose in excess of 150 millirem.

Radiation Burn Radiation damage to the skin; beta burns result from skin contact with or exposure to emitters of beta particles while flash burns result from sudden thermal radiation.

Radiation Chemistry That branch of research concerned with the chemical effects including decomposition of particles on matter.

Radiation Damage General term for the harmful effects of radiation on matter.

Radiation Detection Instruments Devices that detect and record the characteristics of ionizing radiation. *see* COUNTER, DOSIMETER, MONITORING.

Radiation Illness An acute organic disorder that follows exposure to relatively severe doses of ionizing radiation; characterized by nausea, vomiting, diarrhea, blood cell changes, and, in later stages, by hemorrhage and loss of hair.

Radiation Monitoring Continuous or periodic determination of the amount of radiation present in a given area.

Radiation Protection Measures to reduce exposure to radiation.

Radiation Protection Guide The officially determined radiation doses that should not be exceeded, established by the Federal Radiation Council; equivalent to what was formerly termed the MAXIMUM PERMISSIBLE DOSE or EXPOSURE.

Radiation Saturation A phenomenon in which a survey meter's capability to measure radiation levels is overwhelmed, causing the meter to incorrectly read "zero."

Radiation Shielding Reduction of radiation by interposing a shield of absorbing material between any radioactive source and a person or object.

Radiation Source Usually a man-made, sealed source of radioactivity used in teletherapy, radiography, as power source for batteries, or in various types of industrial gauges; machines such as accelerators, radioisotopic generators and natural radionuclides may also be considered as sources.

Radiation Standards Exposure standards, permissible concentrations and rules for safe handling; regulations for transportation and for industrial control of radiation and radiation exposure, set by legislative means.

Radiation Sterilization Use of radiation to cause a plant or animal to become incapable of reproduction; the use of radiation to kill all forms of life, especially bacterias, in food and surgical sutures.

Radiation Therapy Treatment of disease with any type of radiation; radiotherapy.

Radiation Warning Symbol A magenta trefoil on a yellow background is an officially prescribed symbol and should always be displayed when a radiation hazard exists.

Radioactive Exhibiting or pertaining to radioactivity.

Radioactive Cloud Mass of air and vapor in the atmosphere carrying radioactive debris.

Radioactive Dating A technique for measuring the age of an object or sample of material by determining the ratios of various radioisotopes or the products of radioactive decay it contains; example, the ratio of carbon-14 to carbon-12 reveals the approximate age of bones, pieces of wood, or other archeological specimens that contain carbon extracted from the air at the time of their origin.

Radioactive Material (RAM) Any material that spontaneously emits ionizing radiation.

Radioactive Tracer A small quantity of radioactive isotope used to follow biological, chemical or other processes by detection, determination or localization of the radioactivity.

Radioactive Wastes Conventional materials that have been contaminated with radiation; not classified as hazardous and not covered by RCRA, they are specifically controlled by the U.S. Atomic Energy Act.

Radioactivity Spontaneous decay or disintegration of an unstable atomic nucleus accompanied by the emission of radiation.

Radioactivity The analysis of any substance (food, soil, water, etc.) to determine the presence and magnitude of radioactive contamination.

Radiobiology Body of knowledge and the study of the principles, mechanisms, and effects of ionizing radiation on living matter.

Radio Cache (ICS) Portable radios, a base station and repeater stored in predetermined location for dispatch to emergency incidents.

Radio Controlled Engine (RR) An unmanned engine situated within the train consist, separated by cars from the lead unit but controlled from it by radio signals.

Radioecology Study of the effects of radiation on species of plants and animals in natural communities.

Radiography The use of ionizing radiation for the production of shadow images on a photographic emulsion; some of the rays (gamma rays or X-rays) pass through the subject while others are partially or completely absorbed by the more opaque parts of the subject and thus cast a shadow on the photographic film.

Radioisotope An unstable isotope of an element that decays or disintegrates spontaneously, emitting radiation. More than 1300 natural and artificial radioisotopes have been identified.

Radiological A general term referring to processes involving nuclear radiation.

Radiological Emergency Response Plan A detailed program of the actions and responsibilities of local, state, or federal government agencies during a radiological emergency.

Radioluminescence Visible light caused by radiations from radioactive substances; example, the glow from luminous paint.

Radiomutation A permanent, transmissible change in form, quality, or other characteristic of a cell or offspring from the characteristics of its parent, due to radiation exposure.

Radiopharmaceutical A material containing radioisotopes used in medical diagnosis or therapy.

Radium A radioactive metallic element with atomic number 88. An alpha- and gamma-emitter, used as a source of luminescence and as a radiation source in medicine and radiography.

Radius Rod A member used to retain axle alignment, and in some cases control axle torque; can be fixed or adjustable in length.

Radome The cover for an aircraft aerial system which is weatherproof and transparent to radio frequency energy.

Rankine Temperature Scale A scale with the degree interval of the Fahrenheit scale and the zero point at absolute zero; water freezes at 491.60R, boils at 671.69R.

Rate The amount of pesticide (or pesticide formulation) which is being delivered to a plant, animal, or area; measurement of the volume being applied; expressed in gallons per acre, minute or hour or other set measure such as pounds per acre, minute or hour.

Rate of Combustion *see* FLAME SPREAD.

Reaction (1) Chemical: the combining of two or more materials into another, or the chemical breakdown of a chemical material, or the combining of a chemical into a polymer, including oxidation. (2) Physical: the response to a force applied.

Reactive Materials Substances capable of or tending to react chemically with other substances.

Reactivity The degree of ability of one substance to undergo a chemical combination with another substance; the tendency to explode under normal management conditions, to react violently when mixed with water, or to generate toxic gases.

Reactor *see* NUCLEAR REACTOR.

Rear of a Signal (RR) The side of the signal from which the indication is received.

Receiving Facility (EMS) A general acute care (hospital) facility assigned a role in the EMS system by the local EMS agency.

Receiving Track A track used for arriving trains.

Recharge Zone Area through which water enters an aquifer.

Reclamation Restoration to a better or more useful state, such as land reclamation by sanitary landfilling or obtaining useful materials from solid waste; may be used for purposes which are different from their original use.

Recommendation A suggestion from or advice given by an authority.

Reconsignment A service extended by a railroad to the owner of freight permitting a change to the waybill in the name of the shipper, consignee, route or other instructions to effect delivery of the car, providing no back haul is involved.

Recoverable With solid waste, the capability and likelihood of regaining materials for commercial or industrial usage.

Recoverable Resources Materials that still have useful chemical or physical properties after serving a specific purpose and can be reused or recycled for the same or other purposes.

Recovery In oil spills, the entire process of the physical removal of spilled oil from land, water or shoreline environments. General methods of oil recovery from water use mechanical skimmers, sorbents and manual recovery by the cleanup work force; main method of oil recovery spilled on land or shorelines is excavation of contaminated materials.

Recovery Actions Procedures taken after the emergency to restore the affected area as nearly as possible to the pre-emergency condition.

Recovery Period The period of acute radiation syndrome in which the human body recovers from the effects of acute radiation exposure; follows the period of manifest illness.

Recycling A resource recovery method involving collection and treatment of a waste product for use as a raw material in the manufacture of the same or a similar product; also, the marketing of products made from recycled and recyclable materials.

Reduced Speed Prepared to stop short of train or obstruction.

Reduction Removal of oxygen from a compound; lowering of oxidation number resulting from gain of electrons.

Reentry Interval The period of time between a pesticide application and when workers can safely go back into an area without wearing protective clothing or equipment.

Refuse-Derived Fuel (RDF) Fuel produced from solid waste.

References Specific texts or books to consult for information.

Refining in Transit The stopping of shipments of sugar, oil, etc., at a point located between the points of origin and destination to be refined.

Reflector A layer of material immediately surrounding a reactor core that scatters back or reflects into the core many neutrons that would otherwise escape; returned neutrons can then cause more fission and improve the neutron economy of the reactor; common materials are graphite, beryllium, and heavy water.

Reflex Reflector Devices used on side or rear of vehicles to give a warning indication to the driver of an approaching vehicle by reflected light from the headlights of the approaching vehicle.

Refrigeration Charge (RR) A fixed charge for refrigeration from shipping point to destination or for a portion of the trip.

Refrigeration Unit Cargo space cooling equipment.

Refrigerator Car A RR car with insulated walls, floor and roof for carrying commodities that need cooling in transit; two types are those using ice and mechanical means.

Regional An administrative area, division, or district.

Regional Disaster Medical/Health Coordination (RDMHC) (EMS) The individual health officers selected to preplan mutual aid involving mass casualties.

Regional Water Quality Control Boards In California, organizations from each of the nine regions who formulate and adopt water quality control plans for their respective regions and regulate waste discharges from point and non-point sources by establishing and enforcing waste discharge requirements.

Registered Pesticide A pesticide approved by the U.S. Environmental Protection Agency for use as stated on the label of the container.

Registry of Toxic Effects of Chemicals (RTEC) Volumes containing over 58,000 toxicity evaluations of specific chemicals and formulations.

Regular Train Train authorized by a timetable schedule.

Regulated Pest A specific organism considered by a state or federal agency to require regulatory restrictions, regulations, or control procedures in order to protect the host, people and/or the environment.

Regulatory Officials People who work for the federal or state government and enforce rules, regulations, and laws.

Rekindle The process in which a controlled or apparently extinguished fire returns to the free-burning stage; generally caused by improper extinguishment practices and lack of overhaul.

Relative Biological Effectiveness (RBE) Factor used to compare the biological effectiveness of different types of ionizing radiation; inverse ratio of the amount of absorbed radiation required to produce a given effect to a standard radiation required to produce the same effect.

Relative Humidity The percentage of moisture in given volume of air at a given temperature in relation to the amount of moisture the same volume of air would contain at the saturation point.

Relay Emergency Valve A combination valve in an air brake system which controls brake application; also provides for automatic emergency brake application should the trailer become disconnected from towing vehicle.

Release Escape of radioactive materials into the noncontrolled environment.

Release Cock Release valve.

Release Rod (RR) A small iron rod generally located at the side of a car for the purpose of operating the air brake release valve.

Release Valve (RR air brake) A valve attached to the auxiliary reservoir for reducing the air pressure when the locomotive is detached so as to release the brakes.

Relief Valve A valve that will open automatically when pressure gets too high.

REM Radiation Equivalent Man, the unit of dose equivalent; takes into account the effectiveness of different types of radiation.

Remote Assembly Area An area designated outside a hazardous materials incident for the assembly of personnel.

Remote Control Station A station containing equipment to control and regulate operations in an oil field.

Remote Unit *see* RADIO CONTROLLED RR ENGINE.

Remote Sensing Aerial sensing of oil on a water surface; primary applications are the location of an oil spill prior to its detection by any other means, and monitoring of the movement of an oil slick under adverse climatic conditions and during the night.

Removal Frequency and immediacy of clothing removed in case of contamination; information in compliance with government standards or recommendations from recognized authorities.

Repair Track RR track used for repair of cars.

Repeated Contact or Inhalation An exposure to or a breathing in of a pesticide on several occasions over a period of time.

Repellent A pesticide which makes pests leave or avoid a treated area, surface, animal or plant.

Reportable Quantity The minimum quantity of hazardous waste generated as a result of a discharge or spill which must be reported to DER.

Representative Sample A sample of a universe or whole, such as a waste pile, lagoon, or ground water which can be expected to exhibit the average properties of the whole.

Residence Time The time during which radioactive material remains in the atmosphere following the detonation of a nuclear explosive; usually expressed as a half-time, since the time for all material to leave the atmosphere is not well known; half-time.

Residential Waste Waste materials generated in homes and apartments, including paper, cardboard, beverage containers, food cans, plastics, food wastes, glass, garden and yard wastes.

Residual Nuclear Radiation Radiation emitted by radioactive material remaining after a nuclear explosion; arbitrarily designated as that emitted more than one minute after the explosion.

Residual Oils The oil remaining after fractional distillation during petroleum refining; generally includes the bunker fuel oils.

Residual Pesticide A pesticide remaining in the environment for a fairly long time; may continue to be effective for days, weeks, and months.

Residuals Repository A conceptual hazardous waste disposal facility specifically restricted to receiving only residuals from hazardous waste treatment facilities.

Residual Waste Garbage, refuse, or other waste including solid, liquid, semi-solid, or contained gaseous materials resulting from industrial, mining, and agricultural operations and sewage from an industrial, mining or agriculture water supply treatment facility, waste water treatment facility or air pollution control facility, provided it is not hazardous.

Residue Anything remaining after primary purpose has been achieved. (1) Waste materials such as ashes that remain and must be disposed of after gases, liquids or solids have been extracted through resource- or energy-recovery processes. (2) The amount of pesticide remaining on or in a crop or animal or on a surface after it has been treated; the quantity of material remaining after a deposit has aged. Pesticide residues are usually measured in ppm.

Residue Tolerance The amount of pesticide residue permissible in or on a raw agricultural commodity.

Resistant Not affected by a potentially injurious intruder; when an organism is not affected by a pesticide in the manner or to the degree that was expected or that another organism was affected; tolerant.

Resistant Species A species difficult to kill or control with a particular pesticide; one able to suppress or retard the damage usually done by a pesticide.

Resources All of the immediate or supportive assistance available to help control an incident, including equipment, material, and personnel, control agents, agencies, and printed emergency guides.

Resource Conservation Reduction of the amounts of solid waste that are generated; reduction of overall resource consumption and utilization of recovered resources.

Resource Conservation and Recovery Act (RCRA) A federal act giving the Environmental Protection Agency (EPA) authority to develop a nationwide program to regulate hazardous waste from "cradle to grave." Enacted in 1976, the Act was established to "protect human health and the environment from the improper handling of solid waste and encourage resource conservation."

Resource Recovery The extraction and utilization of useful materials (paper, solvents, glass, and metals) or energy from the waste stream either as "raw materials" in the manufacture of new products or as values which can be converted into some form of fuel or energy source; materials that can be reprocessed and reused.

Respiration Breathing; inhalation.

Respirator A face mask which filters out poisonous gases and particles from the air, enabling a person to breathe and work safely; used to protect the nose, mouth, and lungs from hazardous materials.

Respiratory Having to do with breathing (inhalation); of the lungs, nose, mouth and oxygen supply to a human or animal.

Respiratory Toxicity How poisonous a pesticide is to an animal or person when breathed in through the lungs; an intake of any toxic substance through air passages into the lungs.

Response Time (EMS) Interval of time from initiation of an emergency response vehicle to arrival at the scene of the incident.

Response Trust Fund A $1.6 billion fund used for cleanup of abandoned and existing disposal sites; sources of this fund are industrial taxes on oil and certain chemical feedstocks (87%) and federal appropriations (13%).

Restricted Speed (RR) Proceed prepared to stop short of train; obstruction or switch not properly lined, not exceeding 15 mph.

Restricted Use Pesticide A pesticide that has been classified under provision of FIFRA (amended) for use only by an appropriately certified applicator; a limited use pesticide that is applied only by qualified personnel according to the law.

Restrictions Limitations.

Retaining Valve A small, manually positioned valve located near the brake wheel for retaining part of the brake cylinder pressure to aid in retarding the acceleration of a train in descending long grades.

Retarder (RR) A metal grip adjacent to the rails, usually operated by compressed air or electric motors, for regulating speed of a car by pressure on the wheels while rolling down a hump incline.

Retention Time The time hazardous waste is subjected to the combustion zone temperature in an incinerator.

Reuse Multiple use of a product in its original form; differs from recycling in that the product is not reprocessed or refabricated before it is used again.

Revenue Waybill A document showing the amount of charges due on a shipment.

Reverse Lever (RR) The lever which controls the direction of motion of a locomotive by reversing the traction motor field connections.

Richter Scale A measure of the amplitude of earthquake waves at their point of origin, giving an indication of how much energy has been released.

Right-of-Way Property owned by a railroad over which tracks have been laid.

Ringer Solution A water solution containing chemical constituents simulating the blood of animals in which tissues can be kept in a living state for various periods of time.

Ring Stiffener A circumferential tank shell stiffener which helps to maintain the tank cross section.

Rip RR car in need of repair.

Rip Track *see* REPAIR TRACK.

Riser A pipe through which liquid travels upward.

Risk (1) The combination of probability of an accidental occurrence and the likely magnitude of consequences in a given exposure. (2) A situation which may result in death or injury to persons or in damage to property; including effects of toxicity, fire, explosion, shock, concussion, fragmentation, and corrosion.

Risk Analysis A study to determine the threat to a surrounding community.

Risk Assessment Evaluation of the threat to public health and the environment posed by a hazardous waste facility; considering the probability of an incident and its effects.

RNA (Ribonucleic acid) An essential component of all living matter; one form carries genetic information.

Roadbed Foundation on which the railroad track and ballast rest.

Roadside Side of a trailer farthest from the curb when trailer is traveling in a normal forward direction (left hand side); opposite of "curbside."

Roadway *see* RIGHT-OF-WAY.

Rodent Any animal of the order Rodentia; mice, rats, squirrels, gophers, woodchucks.

Rodenticide A pesticide used to control rodents.

Roentgen (R) The unit of radiation exposure in the air; units for quantities of X-ray or gamma radiation measured by detection and survey meters. Named after Wilhelm Roentgen, German scientist who discovered X-rays in 1895.

Roll-Off Truck A vehicle which deposits and collects a 10 to 50 cubic yard container at a site; generally employed in industrial waste collection systems.

Roller Bearings General term applied to a group of journal bearings which depend upon the action of a set of rollers to reduce rotational friction.

Rotary Gauge A variable liquid level gauge consisting of a small position shutoff valve located at the outer end of a tube, the bent inner end resting in the container interior; tube is installed so the tube can be rotated with a pointer on the outside to indicate the relative position of the bent inlet end. The length of the tube and the configuration to which it is bent is suitable for the range of liquid levels to be gaged. The level in the container where the inner end begins to receive liquid can be determined by the pointer position on the outside scale at which a liquid-vapor mixture is observed to be discharged from the valve.

Roughing Filter A prefilter with low efficiency for small particles, usually of the panel type.

Route (1) The course or direction that a shipment moves. (2) To designate the course or direction a shipment shall move.

Rover A joint program of the Atomic Energy Commission and the National Aeronautics and Space Administration to develop a nuclear rocket for space flight.

RPAR Rebuttable Presumption Against Registration. An EPA process to identify pesticide chemicals presenting "unreasonable adverse effects on the environment."

RPM (Revolutions per minute) A measurement of the speed at which something turns or spins.

RTEC Registry of Toxic Effects of Chemicals.

RTEX Number Number assigned by NIOSH to facilitate quick location in the Registry of Toxic Effects of Chemicals.

Runaway Pests Any pest organisms entering a new territory where they have no natural enemies and therefore reproduce with little interference, resulting in a large population which can overrun an area.

Rubbish A term for solid waste that does not include food wastes.

Rule "G" A railroad operating rule prohibiting the use or possession of intoxicants or narcotics while on duty.

Running Gear A general term applied to and including the wheels, axles, springs, axle boxes, frames and other carrying parts of a truck or locomotive.

Running Lights Vehicle marker, clearance, and identification lights required by regulations.

Running Track (1) A railroad track designated in the timetable upon which movements may be made subject to prescribed signals and rules or special instructions. (2) A track reserved for movement through a yard.

Runoff (1) Rainwater, leachate, or other liquid that drains off the surface of the ground. (2) A sprayed liquid (pesticide) which does not remain on the surface of a plant.

Runon Rainwater, leachate, or other liquid that drains overland onto part of a facility.

Runway Defined area prepared for landing and takeoff run of aircraft along its length.

RVR Runway Visual Range; maximum distance in the direction of takeoff or landing at which the runway or the specific lights or markers delineating it can be seen from a position above a specified point on its center line at a height corresponding to the average eye level of pilots at touchdown (approximately 16 feet).

Rupture Disc A safety relief device in the form of a metal disc that closes the relief channel under normal conditions; disc bursts at a set pressure to permit the escape of gas.

S

SAE (Society of Automotive Engineers) A professional organization dedicated to the advancement of the knowledge, experience, and skill in automotive engineering.

Safener A chemical added to a pesticide to keep it from injuring plants.

Safety Chain A secondary connection consisting of chains or cables used between the towing and towed vehicles to retain emergency hook-up in event of separation of the primary towing connection.

Safety Officer Individual on the Command Post staff responsible for the well-being of those operating at a hazardous materials incident; position requires cooperation with many different agencies.

Safety Relief Valve Device on pressure cargo tanks containing an operating part held in place by spring force; valve opens at set pressures.

Safety Rod A standby control rod used to shut down a nuclear reactor rapidly in an emergency.

Salt Compound of the negatively charged ion from an acid and positively charged ion from a metal or alkali base.

Salt water Disposal A system for disposal of salt water produced with crude oil. Typically composed of collection centers where water from several wells is gathered, a central treating plant where the salt water is conditioned, and disposal wells where the treated salt waste is injected into suitable formations.

Salvage The extraction or controlled removal of materials from the solid waste stream for reuse.

Salvage Cover A large vinyl or canvas cover used to protect building contents from water during firefighting operations.

Sample Material containing radionuclides; may consist purely of radionuclides or be a mixture of radionuclides and non-radioactive material.

Sampler Device attached to an oil flow line to permit continuous sampling of the product flowing in the line.

Sanded Up Clogged by sand entering a well bore with the oil.

Sanders Devices operated by air for applying sand to the rail in front of or behind the driving wheels of the railroad engine.

Sand Reel A metal drum on an oil drilling rig around which the sand line (wire rope) is wound.

Sandshoe A flat steel plate serving as ground contact on the supports of a trailer and used instead of wheels, particularly where the ground surface is expected to be soft.

Sanitary Landfill A method of disposing of refuse, usually below ground level, without creating nuisances or hazards to public health and safety.

Sanitary Landfill Site An approved disposal area where materials including garbage, oil-contaminated debris and other solid wastes are spread in layers and covered with soil to a depth which will prevent disturbance or leaching of contaminants towards the surface. Careful preparation of the fill area and control of water drainage are required to assure proper landfilling. Refuse is confined to the smallest practical area and reduced to the smallest practical volume by using heavy tractor-like equipment to spread, compact, and usually cover the waste daily with at least 6 inches of compacted soil. After the area has been completely filled and covered with a final 2 to 3 foot layer of soil and allowed to settle an appropriate period of time, the reclaimed land may be utilized as a recreational area such as a park or golf course; under certain highly controlled conditions the land may be used for some types of building constructions.

Saturated Zone The part of the earth's crust in which virtually all voids are filled with water.

Scale House Structure housing weight recording mechanism used in weighing freight cars.

Scale Test Car A compact RR car equipped with weights for the testing of track scales.

Scale Track A storage track for RR cars needing to be weighed.

Scaler An electronic instrument for rapid counting or radiation-induced pulses from Geiger counters or other radiation detectors; permits rapid counting by reducing (by a definite scaling factor) the number of pulses entering the counter.

Scanner Device used to determine location and amount of radioactive isotopes within the body by measurements taken with instruments outside the body; moves in a regular pattern over area to be studied or over the whole body and makes a visual record.

Scattering A process changing a particle's trajectory, caused by particle collisions with atoms, nuclei, and other particles or by interactions with fields of magnetic force. *see* COMPTON EFFECT.

Scavenging In chemistry, the use of a nonspecific precipitate to remove one or more undesirable radionuclides from solution by absorption or coprecipitation. In atmospheric physics, the removal of radionuclides from the atmosphere by action of rain, snow or dew. *see* FALLOUT.

Schedule That part of a timetable which prescribes class, direction, number, and movement for a regular train.

Scintillation A flash of light produced in a phosphor by an ionizing event. Compare fluorescence, luminescence.

Scintillation Counter An instrument that detects and measures ionizing radiation by counting the light flashes caused by radiation impinging on certain materials (phosphors).

Scram The sudden shutdown of a nuclear reactor, usually by rapid insertion of the safety rods, caused by an emergency or deviation from normal reactor operation.

Scrap Waste material suitable for recovery or reclamation.

Scraper Device used to clean deposits of paraffin from tubing or flow lines (in oil industry).

Scrubber Device using a liquid filter to remove gaseous and liquid pollutants from air stream.

Seals (RR) Metal strips designed for one-time use applied to hasp of closed freight containers; must be broken to remove and are used to indicate that container contents have been undisturbed while in transit; stamped with a name, initial and/or number for identification.

Secondary Materials All types of materials handled by dealers and brokers that have fulfilled their useful function and usually cannot be used further in their present form or at present location, and materials that occur as waste from the manufacturing or conversion of products.

Secondary Track A designated track where trains may be operated without timetable authority, train orders, or block signals.

Secure Landfill One constructed in cell forms to segregate and isolate hazardous materials from each other; clay, plastics or pozzolanic liners form separated, permanent entrapments.

Sediment Material in suspension in air or water; the total dissolved and suspended material transported by a stream or river.

Sedimentation Hazardous waste physical treatment method which separates and removes suspended particles that are heavier than the liquid in which they are present by gravitational settling.

Seebeck Effect Phenomenon involved in operation of a thermocouple; named for German scientist Thomas Seebeck who first observed concept in 1822.

Seed Protectant A chemical applied to seed before planting to protect seeds and new seedlings from diseases and insects.

Selective Pesticide A pesticide more toxic to some types of plants or animals than to others; a selective herbicide might kill crabgrass in a cornfield but would not injure the corn.

Self-Accelerating Decomposition Temperature The temperature above which the decomposition of an unstable material proceeds by itself independently of the external temperature.

Self-Aligning Coupler (RR) A coupler having a taper shank rather than a straight shank to prevent the jackknifing of cars.

Semitrailer A truck trailer equipped with one or more axles and constructed so the front end and a substantial part of its own weight and that of its load rest upon a truck tractor.

Sensitivity Maps Maps used by an on-scene commander and hazardous materials incident response teams which designate areas of biological, social and economic importance in a given region; often prioritize sensitive areas; usually contain other information useful to the response team such as access areas, roads, etc.

Sensitive Easily injured or affected by; susceptible to effect with low exposure or dosage.

Sensitive Areas Locations where pesticide applications could cause great harm; examples: streams, ponds, houses, barns, parks, etc.

Septage Water waste extracted from septic tanks; disposal is the responsibility of State Water Resources Control Board.

Septicemia A disease condition commonly known as blood poisoning, caused by harmful bacteria or their toxic products.

Set-Up Term denoting that an article or area is put together in its complete state; not knocked down.

Sewage Waste flushed from toilets into sewers, then processed in treatment plants. Sewage disposal is responsibility of the State Water Resources Control Board; disposal of sludge or the dried end-product of the sewage treatment process is the responsibility of the State Solid Waste Management Board.

Sheer Section A safety feature in cargo tank piping and fittings designed to fail or break completely to prevent damage to shutoff valves or the tank itself.

Sheet Breakaway In oil spills, a type of current-induced boom failure since a boom placed in moving water acts like a dam; surface water being held back by the boom is diverted downwards and accelerates in attempt to keep up with the water flowing under the boom skirt, thus drawing oil from the surface under the boom.

Shelter A structure or other location offering protection from the elements; in nuclear incidents, a barrier from radiation in the environment.

Shielding Material which can be used to absorb and block radiation energy.

Shift Technical Advisor At a nuclear facility the individual on duty in the Control Room responsible for advising of plant and systems status.

Shifting (RR) Switching.

Shipment Cargo being transported, includes material, its packaging or container, marks, etc.

Shipper Person or firm originating shipment of goods; consignor.

Shipper's Export Declaration Form required by Treasury Department, prepared by consignor showing value, weight, consignee, destination, etc., of shipments to be exported.

Shipper's Load and Count Verification by the consignor that car contents were loaded and counted by the shipper and not the railroad.

Shipping Document A shipping order, bill of lading, manifest, or other document issued by the transport carrier.

Shipping Order, Papers Instructions to the railroad or transport carrier for forwarding all goods; usually the second copy of the Bill of Lading.

Shock Severe reaction of the human body to a serious injury; can result in death if not treated, even if actual injury was not fatal.

Shock Wave A pressure pulse in air, water or earth, propagated from an explosion which has two phases; in the first (positive) the pressure rises sharply to a peak then subsides to the normal pressure of the surrounding medium; in the second (negative) the pressure falls below that of the medium, then returns.

Shoot (1) To explode nitroglycerine or other high explosives in a hole to shatter the rock and increase the flow of oil. (2) In seismographic work, to discharge explosives to create vibrations in the earth's crust.

Shop (RR) Structure where building and repair to equipment is performed.

Shoreline Sensitivity The susceptibility of a shoreline environment to any disturbance which might decrease its stability or result in short or long-term adverse impacts; shorelines most susceptible to damage are usually equally sensitive to cleanup activities which may alter physical habitat or disturb associated flora and fauna.

Shoreline Type The average slope or steepness and predominant substrate composition of a shoreline area.

Short Ton 2,000 pounds.

Short-Term Pesticide A pesticide that breaks down into non-toxic by-products almost immediately after application.

Shredder Mechanical device used to break up waste materials into smaller pieces by tearing and impact actions.

Shroud A panel, similar to a fire wall, separating the engine from the aircraft, pierced by openings for cables, linkage, and plumbing. Following being removed for repairs, seals at the forward bulkhead and forward end of the engine may lose the vapor-tight fit, allowing fuel vapors to collect.

SI The International System of Units; adopted by the General Conference on Weights and Measures, an international diplomatic conference dedicated to developing and unifying the metric system.

SIC Standard Industrial Code. Prepared by the United States Office of Management and Budget.

SIC Number A number assigned to a corresponding type of industry, manufacture, or product under the Standard Industrial Code.

Sidedress Pesticide application along the side of a crop row.

Side Loader A refuse truck in which solid waste is loaded into the side of the vehicle.

Side Marker Lamps Lamps which show to the side of a vehicle; one red lamp mounted as far to the rear as practicable, and one amber lamp mounted as far to the front as practicable, not less than 15 inches above the road surface to show the extreme overall length of the vehicle; an intermediate marker lamp is required on trailers over 30 feet in length mounted midway between the forward and aft marker lamps.

Side Rails The main longitudinal frame members of a tank container used to connect the upper and lower corner fittings.

Side-Track A RR track adjacent to the main track for purposes other than meeting and passing trains.

Siding A RR track adjacent to a main or a secondary track for meeting or passing trains.

Sievert (Sv) The SI unit for measuring dose equivalent; 1Sv = 1joule/kg = 100 rems.

SIGMET Information Aircraft information prepared by a meteorological watch office concerning expected occurrence of adverse weather conditions.

Sign A display used to inform. (1) Some evidence of exposure to a dangerous pesticide; an outward signal of a disease or poisoning in a plant, animal or human. (2) A placard or written display. (3) A hand gesture.

Signal Placard, gesture or lights to convey a message.

Signal Area A selected area of an airport used for display of ground signals to aircraft.

Signal Words and Symbol Words which must appear on pesticide labels in accordance with FIFRA (amended) to show toxicity of that pesticide.

Signal Words	Toxicity	Symbol
Caution	Low	
Warning	Moderate	
Danger-Poison	High	Skull & Cross Bones.

Signboard (RR) Information stencilled on car side pertaining to empty car movement instructions.

Sill The main longitudinal members of a car or truck underframe.

Silt Soil or sediment particles which range in size from 4 to 64 microns; larger than clays (4 microns) and smaller than sand (64 microns to 2mm).

Silvicide A pesticide used to control unwanted brush and trees.

Silviculture Care and cultivation of forest trees.

Simulation A mock accident or release set up to test emergency response methods or for use as a training tool.

Single Resources (ICS) Any resource or grouping of resources on which an individual status is maintained.

Single Track A main RR track upon which trains are operated in both directions.

Sinking Agent Materials spread over the surface of an oil slick to adsorb oil and cause it to sink; rarely used because they may cause considerable damage to bottom-dwelling organisms; include treated sand, fly ash and special types of clay. Provide a purely cosmetic approach to oil spill clean up.

Site (1) A specific location. (2) An area, location, building, structure, plant, animal, or other organism to be treated with a pesticide to protect it from or to reach and control the target pest.

Site Area Emergency An event involving actual or likely major failures of nuclear plant functions needed for the protection of the public; offsite protective actions unlikely.

Situation Unit Leader (ICS) Individual responsible for collection and display of status resources.

Skate (RR) A metal skid placed on rail in a hump yard to stop cars from running out the lower end of the classification yard.

Skew Bridge A bridge crossing a passageway at other than a right angle.

Skid Tank Portable tank, usually positioned on skids or a trailer.

Skin Outer covering of an aircraft.

Skirt Portion of a floating boom which lies below the water surface and provides the basic barrier to the spread of an oil slick or the loss of oil beneath the boom.

Slack Adjuster Adjustable mechanical lever used to transmit brake chamber force to brake cam shaft when the brakes are applied; are designed so they can be adjusted to compensate for lining wear.

Slave Unit (1) A mechanical device directly responsive to another device. (2) A radio controlled RR engine.

Slick Common term used to describe a film of oil on water surface (usually less than 2 microns thick).

Sliding Fifth Wheel A fifth wheel assembly capable of being moved forward or backward on a truck tractor to vary load distribution on the tractor and to adjust the overall length of combination.

Slimicide A pesticide used to prevent slimy growths.

Sling A wire-rope loop for use in lifting heavy equipment.

Slips Wedge-shaped toothed pieces of metal that fit inside a bowl and used to support tubing or other pipe.

Slip Tube Gage A variable liquid level gage with a relatively small positive shutoff valve located at the outside end of a straight tube, normally installed vertically, which communicates with the container interior; tube installation fitting is designed so tube can be slipped in and out of container and the liquid level at the inner end determined by observing when the shutoff valve vents a liquid-vapor mixture.

Slope Sheet Panels located at each end of payload compartment of transport vehicle which direct product by gravity to hoppers.

Slow Board A RR signal indication to proceed at a slow speed.

Slow Speed Not exceeding 15 miles per hour.

Sludge A heavy, slimy solid, semisolid, or liquid waste generated from a municipal, commercial, or industrial waste treatment facility or wastewater treatment plant, waste supply treatment plant or air pollution control facility, exclusive of treated effluent from a wastewater treatment plant.

Slurry Thin mixture of liquid (usually water) and fine particles.

Smoke The dense, opaque gas-like product of a fire; in the chemical sense does not have a fixed composition but consists of many toxic gases, carbon particles and unburned fuel vapors.

Small Quantity Generators Facilities generating or accumulating less than 1,000 kg/month of identified hazardous waste or specific quantities of acutely hazardous wastes; some are exempted from RCRA.

Smear A wipe sample taken of a particular area to determine the presence of loose radioactive contamination.

Snake Out To pull out.

SNAP (Systems for Nuclear Auxiliary Power) An Atomic Energy Commission program to develop small auxiliary power sources for specialized space, land, and sea uses.

SOAR (Save Oil America, Recycle) An oil recycling program directed by the California State Solid Waste Management Board (SSWMB). Program initiated in 1979 as result of SB68, the Used Oil Recycling Act passed by the State Legislature in 1977. More than 2,400 gas stations, stores and recycling centers serve as SOAR collection points where "do-it-yourself" oil changers and others can bring their used motor and industrial oil for recycling. Californians are currently recycling almost half of the used oil generated in the state through the SOAR project.

Soil Application Putting a pesticide in or on the soil instead of applying it directly to vegetables.

Soil Contamination Contamination of the ground area where a pesticide spill or fire occurs, or where contaminated runoff water flows.

Soil Fumigant Pesticide used to control pests in the soil; when added to the soil, takes form of gas or vapor; evaporates quickly, thus often used with some kind of cover (a plastic sheet spread over area) to trap vapor in soil until pest is under control.

Soil Incorporation The mixing of pesticide into soil by mechanical means, usually consisting of pesticide application to the soil followed by some sort of tillage.

Soil Injection Mechanical placement of pesticide below ground level into the soil; usually a pesticide liquid which vaporizes.

Soil Sterilant Chemical prevents growth of all plants and animals in the soil; depending on chemical, may be temporary or long-term.

Sole Use Vehicle A vehicle designated exclusively for transporting radioactive material.

Solidification Process of stabilizing waste into a solid with high structural integrity; solidified wastes are less likely to leach out of land disposal sites than are untreated wastes even though the physical and chemical characteristics of the constituents of the waste may not be changed by the process.

Solid Waste Discarded material other than liquid waste, including paper, dirt, wood, lawn and garden waste, food waste, agricultural waste, industrial waste, demolition and construction waste, glass, ceramics and metal, manure, ash, clothing and rags, leather, rubber and plastic, resulting from industrial, commercial, mining, agricultural operations and from community activities and containing resources that can be recycled, reused, or otherwise recovered.

Solid Waste Disposal Act Amendments of 1980 Amends the Federal Resource Conservation and Recovery Act of 1976 (RCRA) with established effective solid waste management as a national priority. Re: non-hazardous solid wastes, the amendments establish minimum federal funding of state and local agencies for developing solid waste management plans, authorize additional funding for rural community assistance, and authorize State Plans to address energy and materials conservation and recovery. The Solid Waste Disposal Act reauthorizes state surveys of solid waste disposal sites; technological and financial studies of resource-recovery systems by the Environmental Protection Agency (EPA); and development of markets for resource-recovery products by the Secretary of Commerce, as mandated by RCRA. The Used Oil Recycling Act (PL96-483) amended the Solid Waste Disposal Act Amendments to relax labeling restrictions on rerefined oil products and provide funding to the states for used oil recycling programs.

Solid Waste Management The overseeing and regulation of the safe and sanitary reuse or disposal of industrial, residential, construction and agricultural wastes; includes collection, transportation, storage, recycling, resource-reuse and recovery and waste-reduction programs.

Solubility The ability of a substance to mix with water; important in firefighting since flash points and fire points of a flammable liquid are appreciably affected as the amount of water in a mixture is increased.

Soluble Will dissolve in a liquid.

Soluble Powder (SP) Dry preparation containing a fairly high concentration (15%–95%) of active ingredient that dissolves in water or another liquid to form a solution that may be applied.

Solution Mixture of one or more substances into another substance (usually a liquid) in which all ingredients are completely dissolved without their chemical characteristics changing; will not settle out or separate in normal use.

Solvent Liquid which will dissolve a substance to form a solution. Example: water is the solvent when it dissolves sugar. Used in a number of manufacturing/industrial processes including manufacture of paints and coatings for industrial and household purposes, equipment cleanup, and surface degreasing in metal fabricating industries.

Sorbent A substance capable of either adsorbing or absorbing another substance; a natural organic, mineral-based or synthetic organic material used to recover small amounts of spilled material (such as oil).

Sorbent Barrier In oil spills, a boom constructed of or including sorbent materials to simultaneously recover spilled oil during the containment process; used only when the oil slick is relatively thin since recovery efficiency rapidly decreases once the sorbent is saturated with oil.

Sorbent-Surface Skimmer Mechanical skimmer incorporating a rotating sorbent surface (oleophilic) drum, disc, belt or rope to which oil adheres as the surface is moved continuously through the slick.

Source Material which is emitting radiation.

Source Separation The setting aside of recyclable wastes at their point of origin for purposes of recycling; example: home separation of paper, glass and metal from other wastes for separate collection by a recycler.

Source Term A particular type or amount of radionuclide originating at the source of a nuclear incident; also describes conditions and mode of emission.

Space Spray A pesticide applied as tiny droplets which fill the air and destroy insects and other pests, either inside or out-of-doors.

Special Equipment (RR) Freight cars designed to carry specific commodities, some of which contain devices to protect and/or aid in handling shipments.

Special Groups Concentrations of people in one area or building for special circumstances (e.g., schools, hospitals, nursing homes, shopping centers).

Special Information Important data not ordinarily found, such as intricate air measuring ratios or new blood level test criteria.

Special Nuclear Material In atomic energy law, referring to plutonium 239, uranium 233, uranium containing more than the natural abundance of uranium 235 or any material artificially enriched in any of these substances.

Special Protection A means of limiting the temperature of an LP-Gas container for purposes of minimizing the possibility of failure of the container as a result of fire exposure; may consist of applied insulating coatings, molding, burial, water spray fixed systems or fixed monitor nozzles.

Specialized Treatment Facility A facility designed to treat a specific type or group of hazardous wastes, usually generated by a small number of industries with similar production processes.

Species Group of living organisms called by the same common name because they are very much alike and can interbreed successfully.

Specific Activity The rate of decay of a radioactive sample per unit of sample weight; measured in curies per gram. Levels of activity and specific activity are measured with regard to the total weight of radioactive and non-radioactive material in the sample.

Specific Category A pesticide-use designated by FIFRA or by state regulations; a special area (such as Forest Pest Control or Aquatic Pest Control) requiring state certification for use of restricted materials by commercial applicators.

Specific Gravity Ratio of the weight of the volume of liquid or solid to the weight of an equal volume of water; specific gravity of water is 1.0; material with a specific gravity below 1.0 will float on water, that with a specific gravity over 1.0 will sink in water.

Specific Heat Measurement of the ability of a substance to absorb heat, indicated by the relative quantity of heat needed to raise the temperature of one pound of the substance one degree Fahrenheit; water has highest specific heat of any known substance.

Specific Pesticide Same as SELECTIVE PESTICIDE.

Spectrum A visual display, photographic record, or plot of the distribution of the intensity of a given type of radiation as a function of its wave length, energy, frequency, momentum, mass or any related quantity.

Spent (Depleted) Fuel Nuclear reactor fuel that has been used to the extent that it can no longer effectively sustain a chain reaction.

Spill The accidental release of contaminating materials.

Spillage Any escape, leakage, dripping, or running over of materials.

Splash Guard Deflecting shield sometimes installed on tank trailers to protect meters, valves, etc.

Splice Bar *see* JOINT BAR.

Splitter Valve Valve installation to divide pipeline manifold.

Spontaneous Fission Fission occuring without an external stimulus; process occurs occasionally in all fissionable materials.

Spontaneous Ignition A material proceeding without constraint by internal impulse or outside energy to kindle or set fire; quick or slow oxidation or combustion brought about by chemical, electrical, biological (bacterial) or physical processes (vibration, pressure, friction) without assistance of extraneous sources of heat (flame, sparks, hot or glowing bodies).

Spontaneously Combustible The process of increase in temperature of a material to a point of ignition without drawing heat from its surroundings.

Spot Place a RR car in a designated position or specific location usually for loading or unloading such as at a warehouse door.

Spot System A classification system for repair work where RR cars are moved progressively from one spot to another with different items of work done by a unit force.

Spot Treatment Pesticide application directed at small area such as at specific plants; opposite of General Treatment.

Spray Drift Same as DRIFT.

Spread Tandem A two-axle assembly wherein the axles are spaced to allow maximum axle loads under existing regulations; distance between center of axles of a spread tandem generally has been more than 50″.

Spreader An adjuvant which increases the area that a given volume of liquid will cover on a solid surface.

Spreader Bars Transverse rigid stabilizing rods attached to the side walls of a trailer; common usage is on open top van trailers.

Spring Deflection Depression of a trailer suspension spring when the spring is placed under load.

Spring Seat Suspension component used to support and locate spring on axle; formerly called spring chair.

Spring Switch RR switch equipped with a spring mechanism arranged to restore the switch points to normal position after having been trailed through.

Spur Track A stub RR track extending out from a main track.

Stabilization (1) Stage of an incident when the immediate problem or emergency has been controlled, contained, or extinguished. (2) Hazardous waste chemical treatment method by which a chemical reaction produces an insoluble form of the waste or incorporates the waste into a form that is insoluble.

Stabilization Lagoon A shallow pond for the storage of waste water before discharge to a stream or to a treatment facility.

Stabilizer Inhibitor.

Stabilizing Jack A device usually of single leg construction used on the forward end of a trailer as an auxiliary support to prevent nosediving.

Stable (1) Incapable of spontaneous change. (2) Not radioactive.

Stage of the Incident One of five definite, identifiable phases through which an emergency passes from onset (interruption of normal conditions) to stabilization.

Staging Area (ICS) Location where incident personnel and equipment are assigned in anticipation of immediate assignment to combat operations.

Standard Gage Distance of 4′ 8½″ between the heads of the rails.

Standard Operating Procedures Detailed instructions for implementation of emergency plans by the various response agencies.

Standard Rate Rate established via direct routes from one point to another.

Standard Transportation Commodity Code (STCC Code Number) A listing of code numbers for categories of articles being shipped, in general use by carriers.

State Law Abstracts of state laws, regulations, and standards concerning hazardous substances.

State Override Refers to procedures whereby local decisions on hazardous waste management facility siting can be overturned by state agency.

State Preemption Refers to the preemption of local decision-making authority over hazardous waste management facility siting by state agency so that no local decision is required to site such facilities.

State Siting Regulations Agency-issued, state authorized directives to implement RCRA dealing with the approval or restriction of facility locations and/or permitting processes.

State Water Resources Control Board (SWRCB) The State Board, established by the Porter-Cologne Water Quality Control Act of 1967, adopted the California state policy for water quality control in accordance with legislative policies, and is designated as the state water pollution control agency for the purpose of the Federal Water Pollution Control Act; reviews actions of the Regional Water Quality Control Boards relative to regulation of waste discharges.

Stationary Installation (Fixed or Permanent Installation) An installation of LP-Gas containers, piping and equipment for use indefinitely at a particular location; an installation not normally expected to change in status, condition or place.

Statistically Significant When the difference between a predicted and an observed value is so large that it is improbable that it could be attributed to chance.

Status Under Law Lists all federal regulations, rules, and standards,

NIOSH recommendations, and agency lists for scientific evaluation and potential regulation.

Stay Time Period of time personnel may work within a radiation area before receiving a dose equal to the field administrative limit; 25 rems recommended.

Steel Any of various hard, strong, durable, malleable alloys of iron and carbon, usually containing between 0.2 and 0.7% carbon, often with other constituents such as manganese, nickel, copper, tungsten, cobalt, or silicon, and widely used as a structural material.

Step-Off Area Area designated for entry to and exit from a radiation area.

Sticker An adjuvant which increases the adherence of a pesticide.

Stock Car A car for the transportation of livestock equipped with slatted sides, single or double deck and sometimes with feed and water troughs.

Stolon An above-ground stem that produces roots.

Stomach Poison A pesticide which kills an animal when eaten or swallowed.

Stopping in Transit The holding of a shipment by a carrier on order of the owner after the transportation movement has started and before it is completed.

Stopping Power A measure of the effect of a substance upon the kinetic energy of a charged particle passing through it. *see* ABSORPTION.

Storage Containment of waste on a temporary basis in such a manner as not to constitute disposal of such waste; containment of such waste for over a year constitutes disposal.

Storage Facility Any location storing wastes for less than 90 days for subsequent transport off-site (except generators who store their own wastes).

Storage Tank Any manufactured non-portable covered device used for containing pumpable hazardous wastes.

Storage in Transit The stopping of freight at a point located between the point of origin and destination to be stored and forwarded at a later date.

Storage Track A RR track on which cars are placed when not in service.

Streaming The channeling of radiation into a beam; this condition would occur if the lid of a shielding cask were removed, allowing radiation to escape the container; the radiation would be in the form of a column, similar to the beam of a flashlight.

Stress (1) (Verb) To apply a form of force or energy that tends to strain or deform or otherwise change a body or mass. (2) (Noun) The condition of being subjected to force or energy tending to deform or strain. (3) A state of tension put on or in a shipping container by internal chemical action, external mechanical damage, or external flames or heat.

Stressee The actor whose state is being changed by stresses.

Stressor The actor applying the force or energy being transferred.

String Two or more RR freight cars coupled together.

Structures The basic installations and facilities on which a community depends (roads, powerplants, utilities, communication systems).

Stub Track A track connected at one end only.

Subacute Toxicity How poisonous a pesticide is to an animal or person after repeated doses (long-term exposure).

Subcritical Mass An amount of fissionable material insufficient in quantity or of improper geometry to sustain a fission chain reaction.

Subcutaneous Beneath the skin. In toxicological sense, the introduction of a substance by injection directly beneath the dermis.

Subframe Assembly employed to unit trailer body and suspension.

Sublimination Changing of a substance from a solid to a gaseous state without ever going through the liquid phase.

Submersion Skimmer In oil operations, a type of mechanical skimmer incorporating a moving belt inclined at an angle to the water surface in such a way that oil in the path of the device is forced beneath the surface and subsequently rises (due to its buoyancy) into a collection well.

Subsea Completion System A submersible apparatus similar to a bathysphere in which men are lowered to the ocean bottom to work on the wellheads of completed oil wells. The wellhead contains a cellar to which the system is attached, allowing a dry work area.

Substrate Materials which form the base of something. In biology, the base on which an organism lives. Substrate materials include water, soils, and rocks as well as other plants and animals.

Suction Line The line carrying a product out of a tank to the suction side of a pump.

Suction Skimmer A type of mechanical skimmer incorporating an enlarged intake device at the end of a vacuum hose to increase the surface area over which suction of an external pump is exerted.

Suggested Control Measures Procedures necessary to attain federal and state ambient air quality standards.

Suit in Equity To ask the courts to order someone to do something besides pay money. DER has the authority to ask a court to order a hazardous waste handler to do something which it is supposed to do, or to not do something (e.g. pollute).

Sump (1) The low point of a tank at which the emergency valve or outlet valve is attached. (2) A stationary device designed to contain an accumulation of hazardous waste resulting from a hazardous discharge from a tank, container, waste pile, surface impoundment, landfill, or other hazardous waste management structure.

"Superfund" The Comprehensive Environmental Response, Compensation and Liability Act of 1980 provides the federal government with the mechanism to take emergency or remedial action to clean up both abandoned and existing disposal sites whenever there is a release or potential threat of a release of a hazardous substance which may present imminent and substantial danger to public health and welfare; funds for these cleanup actions come from a $1.6 billion trust fund called the "Response Trust Fund."

Superheating Heating of a vapor, particularly saturated steam to a temperature much higher than the boiling point at the existing pressure; occurs in power plants to improve efficiency and to reduce condensation in the turbines.

Supertanker A tanker with a capacity over 100,000 deadweight tons.

Superior Train A train having precedence over another train.

Supplement Any substance added to a pesticide to improve its performance. *see* ADJUVANT.

Supplies Equipment and all expendable items assigned to an incident.

Supply Officer The individual in a Command Post responsible for movement, requisitioning, and issuing of materials, clothing or equipment to be used at the scene of an emergency incident. Also: Supply Unit Leader.

Supportive Care (EMS) Measures taken to maintain vital body functions.

Supports Devices generally adjustable in height, used to support the front end of a semitrailer in an approximately level position when disconnected from towing vehicle.

Surety Bond A guarantee of performance to complete a contract or obligation.

Surface Active Agents Chemicals which alter the forces of surface tension between adjacent molecules; generally decrease the surface tension of a fluid such as an oil, used to facilitate its dispersion throughout the water column.

Surface Impoundment Facility or part of a facility which is a natural topographic depression, man-made excavation or diked area formed primarily of earthen materials (although may be lined with synthetic materials) designed to hold an accumulation of liquid wastes or wastes containing free liquids, and which is not an injection well. Examples: holding, storage, settling, and aeration pits, ponds, and lagoons.

Surface Spray A pesticide sprayed evenly over the entire outside of the object to be protected.

Surface Tension Force of attraction between the surface molecules of liquid; affects the rate at which a spilled liquid will spread over a land or water surface into the ground.

Surface Water Water located above ground, such as ponds, lakes, streams, rivers, etc.

Surfactant An adjuvant which improves the emulsifying, dispersing, spreading, and wetting properties of a pesticide.

Surveillance Periodic medical examinations required by OSHA standards; listed by individual medical tests and procedures.

Survey Meter An instrument designed to detect and measure radiation.

Susceptibility Degree to which an organism can be injured or affected by a pesticide at a known dosage (or exposure).

Susceptible Capable of being injured, diseased, or poisoned.

Susceptible Species A plant or animal that is poisoned by moderate amounts of a pesticide.

Suspended (1) Dispersed in a liquid. (2) A pesticide use no longer legal, where the remaining stocks cannot be used (more severe than Cancelled).

Suspension (1) An assembly employed to connect axle to subframe in such a manner as to cushion road shock and to carry the load. (2) Combination of a substance with a liquid in which the substance is not dissolved. (3) Pesticide formulation in which finely divided solid particles of an active ingredient are mixed in a liquid.

Swab Device that fits the inside of tubing closely and is pulled through to clean the tubing, or to pull such a device through the tubing.

Swamper Helper on a truck.

Swath The width of ground covered by a sprayer when it moves across a field or other treated area.

Switch (1) Connection between two lines of RR track to permit cars or trains to pass from one track to the other. (2) To move RR cars from one place to another within a yard, industry, or terminal.

Switch Back A series of zigzag curves in mountainous terrain to reduce the rate of climb or descent.

Switch Engine Locomotive used for moving cars in terminals and yards; usually built to carry all its weight on the driving wheels.

Switch List (RR) A list of freight cars in track standing order showing cars by initial, number, type of car, and showing where cars are to be switched as required by local practice.

Switch Lock A fastener, usually a spring padlock, used to secure the switch or derail stand in place.

Switch Order Instructions to move a RR car from one place to another within switching limits.

Switch Stand Device by which a switch is thrown, locked and its position indicated; consists of base, spindle, lever, and connecting rod and usually has a lamp and a banner signal.

Switch Target A visual day signal fixed on the spindle of a switch stand, or the circular flaring collar fitted around the switch lamp lens, and painted a distinctive color to indicate the position of the switch.

Switch Tender Employee responsible for aligning tracks for engine and car movements by throwing switches.

Symptom A warning that something is wrong; a feeling of being sick; an indication or poisoning or disease in a person, animal, or plant.

Synapse The point of contact between the axon of one neuron and the cell body or dendrites of another nerve fiber ending where biochemical reactions occur in nerve transmission.

Syndrome A series of symptoms associated with a specific illness.

Synecology The study of ecology dealing with interrelationships of living communities of organisms to each other and to the environment.

Synergistic When the combined action of two or more pesticides is greater than the sum of their activity when used alone.

Synonyms Common name, slang name, acronym, trade name, brand name, or alternate chemical name.

Synthetic Organic Pesticides Man-made pesticides which contain carbon, hydrogen, and other elements.

Synthetic Organic Sorbent One of several organic polymers, genally in the form of plastic foams or plastic fibers, used to recover spilled oil; have higher recovery capacities than either natural organic sorbents or mineral-based sorbents and many can be reused after oil is squeezed out. Examples: polyurethane foam, polyethylene and polypropylene.

System Organization and Management Element (EMS) The structure of an EMS system; the administration, policy making, planning, data collection and analysis, evaluation, and financing.

Systemic A pesticide absorbed by one part of a plant or animal and moved to another section where it acts against a pest. Example: A systemic insecticide can be applied to the soil, be absorbed by the plant's roots, move into the leaves, and then control insects when they feed on the leaves.

Systemic Insecticide That pesticide absorbed into the plant or animal to be protected, in order to control the attacking pest.

T

Tactics Successful methods or procedures used to deploy various resources to achieve objectives.

Take Internally Ingest; to eat or swallow; to take by mouth into the digestive system.

Tamper A power-driven machine for compacting ballast.

Tandem Two-axle suspension.

Tank (1) A large receptacle for holding, transporting, or storing liquids. (2) A stationary device designed to contain an accumulation of hazardous waste, constructed primarily or entirely of nonearthen materials such as concrete, steel, or plastic which provides structural support and containment.

Tank Car A RR car used for carrying liquids, such as oil, molasses, vinegar, acid, etc.

Tank Dome A vertical cylinder attached to the top of a tank car, permitting tank proper to be filled to full cubical capacity, allowing room for expansion of the contents into the dome.

Tank Table A chart giving the barrels of fluid contained in a storage tank corresponding to the linear measurement on a gauge line. Prepared from tank strapping measurements.

Tap Line (RR slang) A short railroad usually owned or controlled by the industries which it serves and connecting with a trunk line.

Tar A black or brown hydrocarbon material ranging in consistency from a heavy liquid to a solid; most common source of tar is the residue left after fractional distillation of crude oil.

Tar Balls Compact semi-solid or solid masses of highly weathered oil formed through the aggregation of viscous, high carbon-number hydrocarbons with debris present in the water column; generally sink to the sea bottom but may be deposited on shorelines where they tend to resist further weathering.

Tare Weight (1) Weight of a container and the material used for packing. (2) Weight of any empty freight car.

Target An area, building, animal, plant, or pest which is to be treated with a pesticide.

Target Pest The pest at which a pesticide application or other control method is directed.

Task Force (ICS) A group of suppression and rescue resources temporarily assembled for a specific mission.

TCA Terminal Control Area; all air space surrounding an airport in which aircraft must be in radio contact with the Air Traffic Controller, approximately 50 miles diameter and 8000 feet elevation.

Team Track A RR track on which cars are placed for public use to load or unload freight.

Technical Assistance Personnel, agencies, or printed materials providing technical information on the handling of hazardous materials.

Technical Assistance Grants Funds made available to communities to hire technical experts to assist in the review of hazardous waste management facility proposals.

Technical Material In pesticides, the active ingredient as it is manufactured by a chemical company before formulation.

Technical Pesticide Highly concentrated pesticide which is to be combined with other materials to formulate pesticidal products.

Technical Specialist Individual or unit responsible for the collection, evaluation and dissemination of information concerning specialized data.

Telecommunications Pertaining to the transmission of signals over long distances, such as by telegraph, radio, or television.

Telemetry (1) System by which the variations recorded by any physical or other instrument can be shown at a distance by means of electricity. (2) (EMS) Transmission of digital and analog biomedical data.

Temperature A condition of hotness or coldness whose state is being described and is relative only to the condition being observed. Measured artificially by scales of different types, i.e., Fahrenheit, Centigrade, Kelvin, Rankin, etc. Absence of heat is manifested as cold temperatures; presence of heat is referred to as high temperatures.

Tenacity The resistance of a deposit to removal by weathering.

Tension Member The part of a floating containment boom which carries the load placed on the barrier by wind, wave, and current forces; commonly constructed from wire cable due to its strength and stretch resistance.

Teratogen A substance which may cause abnormal development in unborn animals.

Teratogenesis Alteration in the formation of cells, tissues, and organs resulting from physiologic and biochemical changes in the fetus during growth; may affect function as well as structure of developing cells (occurs very early in fetal period).

Terminal Carrier The railroad making delivery of a shipment at its destination.

Terminal Charge (RR) A charge made for services performed at terminals.

Terminals (1) Points where employees in road train and engine service originate and/or terminate their tour of duty. (2) A device attached to the end of a wire or cable or to an electrical apparatus for convenience in making connections.

Terrain Clearance Warning Indicator An airborn radionavigational device giving a warning signal when aircraft clearance from the ground immediately below reaches a pre-determined minimum value; a low altitude warning.

Test Animals Laboratory animals exposed to pesticides so that toxicity and hazards can be determined.

Test Weight Car (RR) Scale Test Car.

Tests Medical diagnoses used as an index of exposure to a specific substance.

T-Handle Aircraft emergency power shutdown device, usually located on main instrument panel or overhead panel.

Therapeutic Care (EMS) Measures taken to relieve or cure bodily dysfunctions resulting from injury or illness.

Thermal Of, about, or related to heat.

Thermal Treatment Process by which hazardous waste is rendered nonhazardous or is reduced in volume by exposing the waste to high temperatures; organic materials are oxidized and converted to carbon dioxide and water.

Thermoluminescent Dosimeter (TLD) A device used for measuring beta, gamma, and X-ray radiation exposure; energy absorbed from the radiation raises the molecules of the material in the TLD to excited states where it remains until it is heated; at that time it gives off an amount of light proportional to the dose and that light is measured with a photomultiplier tube.

Thermometer A device for measuring heat intensity.

Thermonuclear Reaction A process where very high temperatures bring about the fusion of two light nuclei to form the nucleus of a heavier atom, releasing a large amount of energy; in a hydrogen bomb, the high temperature to initiate the thermonuclear reaction is produced by a preliminary fission reaction.

Third Rail (RR) An electric conductor located alongside the running rail from which power is collected by means of a sliding contact shoe attached to the truck of electric equipment.

Third Rail Shoe (RR electric locomotive) An insulated metallic sliding contact, mounted on the truck of an electric locomotive for collecting current from an insulated third-rail located alongside the running rails. Positive contact between shoe and rail is maintained by gravity, a spring, or by pneumatic pressure.

Thorium A naturally radioactive element with atomic number 90 and, as found in nature, an atomic weight of approximately 232. The fertile thorium 232 isotope is abundant and can be transmitted to fissionable uranium 233 by neutron irradiation.

Threshold The point where a physiological or toxicological effect begins to be produced by the smallest degree of stimulation.

Threshold Limit Value The concentration of a toxic substance which can be tolerated with no ill effects.

Thrombocytes Platelets, or blood components which are needed in the clotting process.

Through Rate (RR) A rate applicable from point of origin to destination; may be either a joint rate or a combination of two or more rates.

Through Train Train not stopping at all stations on its route.

Thyroid Blocking Agent Non-radioactive iodine or iodine substitute administered to limit the uptake of ingested or inhaled radioiodine by the body; a thyroid phophylaxis.

Tide Pools Permanent depressions in the substrate of intertidal zones which always contain water but are periodically flushed with successive incoming tides; most frequently located near high tide mark, often contain abundant flora and fauna.

Tie Down Device to which a guy wire or brace may be attached.

Tie 'Em Down (RR slang) Apply handbrakes.

Tie On (RR) Couple on.

Tie Plate A metal plate at least 6 inches wide and long enough to provide a safe bearing area on the railroad tie, with a shoulder to restrain outward movement of the rail.

Time Interval Period required between the final pesticide application and harvesting; ensures the legal residue tolerance will not be exceeded.

Timetable Authority for movement of regular trains subject to the rules; contains classified schedules with special instructions relating to the movement of trains and engines.

Tin A silvery metallic element used to coat other metals to prevent corrosion and form parts of numerous alloys.

Tin Can Essentially a steel can with a tin (approximately .0015″) coating; this tin represents one-third of the recycled value of the can while comprising only 0.25% to 4.0% by weight.

Tin Hat Hard hat; metal hat worn by construction and oil-field workers to protect them from falling objects.

TIR Transport International Routier, translated: International Transport by Road. Customs agreement existing among many countries, principally European, permitting vehicles or containers (properly approved and certificated) to be sealed under Customs direction in one country and be transported across borders of member countries without reinspection until arrival at final destination where seal is removed under Customs supervision.

Tire Carrier Spare tire rack.

Tire Clearance Distance between tires and the nearest parts of the trailer body and underconstruction.

TLD Thermoluminescent Dosimeter.

TNT Trinitrotoluene, an explosive.

TNT Equivalent A measure of energy released in detonation of a nuclear explosive expressed in terms of the weight of TNT which would release the same amount of energy when exploded; usually expressed in kilotons or megatons. The TNT equivalence relationship based on fact that 1 ton TNT releases one billion calories of energy.

TOFC (RR) Trailer on Flat Car; Piggy-back.

Tolerance Ability to withstand unfavorable conditions. In use of pesticides, the amount of pesticide that legally and safely may remain in or on any raw farm products at the time of sale (if the feed or plant or animal is to be eaten by livestock or people).

Tolerant Not susceptible to a pesticide application; resistant.

Ton-Mile (1) Movement of one ton of freight one mile. (2) Unit used in comparing freight earnings or expenses; the amount earned from or the cost of hauling a ton of freight one mile.

Tool Train (RR slang) Wreck train used to clear derailments.

Topical Application An application of pesticide from above, usually to the top or upper surface of a plant.

Topography Configuration of the surface area including its relative elevations and the position of natural and artificial features.

Torching Burning with a white flame; the burning of explosives.

Total Elapsed Time Total time period; from point that heat source raises a fuel to ignition temperature and that fuel begins to ignite to the moment that the combustion process stops, whether as result of fire department operations, lack of oxygen or lack of fuel.

Totally Enclosed Treatment Facility A hazardous treatment facility directly connected to an industrial production process, constructed and operated to prevent waste release into the environment.

Tow Bar A beam structure used to maintain rigidly the distance between a towed vehicle and the towing vehicle.

Tower Building of sufficient height to permit maximum viewing. (RR) House for yardmaster, switch lever operator, block operator, or dispatcher.

Toxic Poisonous; relating to or caused by toxin; able to cause injury by contact or systemic action to plants, animals or people.

Toxicant Poison; an agent capable of being toxic.

Toxicity The degree of being poisonous; capability of a poisonous compound to produce deleterious effects in organisms such as alteration to behavioral patterns or biological productivity or death.

Toxicology The study of chemical substances which exert deleterious effects on living organisms, their chemistry in relation to their mode of action, antidotes, and physiological effects.

Toxic Waste Refuse posing a substantial present or potential hazard to human health or to the environment when improperly managed; includes poisonous wastes, carcinogenic, mutagenic, teratogenic, phytotoxic or toxic to aquatic species.

Toxin A poison produced by a plant or animal.

Tracer (1) To discover by going backwards over the trail of (something). (2) With freight, to search for a shipment to expedite movement or establish delivery.

Track (RR) The space between the rails and space of not less than four feet outside each rail.

Track Car (RR) A self-propelled car which may not operate signals or shunt track circuits; includes burro cranes, detector cars, weed burners, tie tampers, etc.

Track Circuit An electrical circuit including the rails and wheels of the train; used for controlling signal devices (fixed signals as well as flashers and gates at crossings).

Trade Name Brand name.

Traffic Control System A block signal system under which train movements are authorized by block signals, whose indications supersede the superiority of trains for both opposing and following movements on the same track.

Trailing Movement Movement of a train over points of a switch which face in the direction in which the train is moving.

Trailing Point Switch A switch on which the points face away from approaching traffic.

Train An engine or more than one engine coupled, with or without cars, displaying a marker.

Train Line The complete line of air brake pipes in a train, including the rigid piping secured under cars and flexible connections between cars and the locomotive.

Train of Superior Direction Train given precedence in the direction specified by timetable as between opposing trains same class.

Train of Superior Right Train given precedence by timetable.

Transfer Agreement (EMS) Written contract between health facilities providing reasonable assurance concerning transfer of patients between facilities when medically appropriate as determined by the attending physician.

Transfer Station A facility designed to accept wastes from the surrounding region for purpose of temporary storage and/or repackaging. When wastes have been collected in sufficient quantity to make shipment economically practical, they are transported in larger trucks in a compact and orderly manner to disposal site or resource/recovery facility.

Transformation Changing from one form to another; transmutation.

Transient Rapid change of a plant operating parameter such as temperature, pressure, or steam generator water level.

Transition That section of a tank joining two unequal cross sections.

Transit Privilege Service granted to a shipment en route such as milling, compressing, refining, etc.

Transit Rate A rate restricted in its application to traffic which has been or will be milled, stored or otherwise specially treated in transit.

Translocated Pesticide A pesticide that moves within a plant or animal after it has entered by some path; a systemic pesticide.

Transmitter Apparatus for production and modulation of radio frequency energy for purpose of radio communication.

Transmutation The changing of one element into another by a nuclear reaction or series of reactions. Example: the transmutation of uranium-238 into plutonium-239 by absorption of a neutron.

Transport To carry from one location to another.

Transport Mode Method of transportation (highway, rail, water, pipelines, air).

Transport Time (EMS) The interval of time required for emergency medical transport of patient from the scene of an incident to a receiving facility.

Transportation Component (EMS) Combined personnel, equipment resources and arrangements to deliver initial emergency medical care to the location of a patient, can rescue the patient from a hazardous environmental situation, and then transport that patient to the appropriate medical facility.

Transposing The act of rearranging numbers or letters from their proper order.

Trash Waste materials usually not including garbage; may include other organic materials such as plant trimmings.

Treated Area Location where a pesticide application has been made.

Treatment Any method, technique, or process which changes the physical, chemical or biological composition of any hazardous waste and so renders it non-hazardous, safer for transport, capable of recovery and/or storage, or reduces its volume.

Treatment Facility Any facility where hazardous waste is subjected to treatment or where a resource is recovered from hazardous waste.

Treatment Zone The area within a land treatment unit where all degradation, transformation, or immobilization of hazardous constituents must occur.

Triage Process of determining which casualties need urgent treatment, which are well enough to go untreated, and which are beyond hope of benefit from treatment.

Tri-Axle Three-axle-construction in which at least two axles are equally spaced approximately 48–50″ apart; the third may be spread or equally spaced.

Trilevel Car A three level freight car used for transporting automobiles.

Triple Load (RR) Shipment requiring use of two carrying cars with an idler car between.

Tritium A radioactive isotope of hydrogen with two neutrons and one proton in the nucleus; synthetic and heavier than deuterium (heavy hydrogen), it is used in industrial thickness gages and as a label in experiments in chemistry and biology; nucleus is a triton.

Truck A self-propelled vehicle carrying its load on its own wheels and primarily designed for transportation of property rather than passengers.

Truck Tractor A powered motor vehicle designed primarily for drawing semitrailers, and so constructed as to carry part of the trailer weight and load.

Truck Trailer A vehicle without motor power, primarily designed for transportation of property rather than passengers, and to be drawn by a truck tractor.

Truck Trailer Manufacturers Association (TTMA) National trade organization representing the manufacture of all types of commercial, non-powered, property-carrying truck trailers, containers and container chassis, and the components of each.

TSDFs (Hazardous waste) treatment, storage and disposal facilities.

Turn Signal Lamps Lights mounted on right, left, front and rear sides of a vehicle which indicate a change in direction by giving flashing warning signals on the side toward which the turn will be made.

Twist Lock A mechanically-operated device located on corners of a container chassis and on automatic lifting spreaders; used for restraining container during transport or transfer.

Type A Packaging Packaging designed in accordance with general packaging requirements for normal commercial shipments.

Type B Packaging Packaging designed in accordance with the general packaging requirements to withstand accident conditions and still remain intact.

U

UHF (Ultra High Frequency) Radio spectrum between 300–3000 MHz. Includes EMS channels from 450:470 MHz.

Ultra-Low Volume (ULV) Spray application of a pesticide that is almost pure active ingredient, sprayed over a large area.

Ultraviolet Radiation The portion of the electromagnetic spectrum emitted by the sun adjacent to the violet end of the visible light range. Often called "black light," it is invisible to the human eye but when it falls on certain surfaces it causes them to fluoresce or emit visible light; responsible for the photo-oxidation of certain compounds including hydrocarbons.

Unassigned Car RR car, usually with some interior loading devices, but not assigned to a particular industry or commodity.

Unclaimed Freight Freight shipment which has not been called for by consignee or owner.

Uncoupling Lever, Uncoupling Rod (RR) A device to uncouple RR cars without moving between them, this rod has a bent handle forming a lever and is usually attached to the end sill; lever proper is attached to the rod and operates the unlocking mechanism, but in the case of freight cars the lever and rod are generally constructed in one piece.

Undercarriage Complete subframe, suspension with one or more axles which may be interconnected, and wheels, tires, and brakes.

Underground Water Water and waterways located below soil surface.

Uniform Coverage Even application of pesticide over an entire area, plant, or animal.

Uniform Demurrage Rules Schedules providing rules and charges for demurrage which are, in general, used throughout the United States, having the approval of but not prescribed by the ICC.

Uniform Freight Classification A listing of commodities showing their assigned class rating to be used in determining freight rates, together with governing rules and regulations.

Unit (ICS) That organizational element having functional responsibility for a specific incident planning, logistic or finance activity.

UN United Nations.

Universal Cylinder A DOT cylinder specification container, constructed and fitted with appurtenances in such a manner that it may be connected for service with its longitudinal axis in either vertical or horizontal position, and so that its fixed maximum liquid level gage, relief device(s) and withdrawal appurtenance will function properly in either position.

Unsaturated Zone Region between the land surface and the upper boundary of the zone of saturation or water table (zone of aeration).

Unstable Ready to behave erratically or in an unwanted manner.

Unstable Materials Substances capable of rapidly undergoing chemical changes or decomposition.

Unusual Event Off-normal events which do not by themselves constitute significant events at a nuclear facility, but could indicate a potential degradation in the level of safety.

Upper Coupler Assembly An upper coupler plate, reinforcement framing and fifth wheel kingpin mounted on a semitrailer; formerly called upper fifth wheel assembly.

Uranium The basic raw material of nuclear energy, uranium is a radioactive element with the atomic number 92 and, as found in natural ores, an average atomic weight of approximately 238. The two principal natural isotopes are uranium 235 (0.7% of natural uranium) which is fissionable and uranium 238 (99.3% of natural uranium) which is fertile. Natural uranium also includes a minute amount of uranium 234.

Uranium Hexafluoride A volatile compound of uranium and fluorine; the process fluid in the gaseous diffusion process.

Uranium Tetrafluoride A solid green compound called green salt; an intermediate product in the production of uranium hexafluoride.

Uranium Trioxide Orange Oxide; an intermediate product in the refining of uranium.

Used Motor Oil Any oil previously used in any machinery; main markets are in road oiling, industrial fuel, and rerefining.

USDA United States Department of Agriculture.

V

Vacuum Space devoid of all matter and exerting zero pressure.

Vapor Gases given off, with or without the aid of heat, by substances that under ordinary circumstances are either a solid or a liquid; gas, steam, mist, fog, fume, or smoke.

Vapor Density Measurement of the weight of vapor as compared with an equal volume of dry air; a figure of less than 1 indicates a vapor lighter than air which will rise; vapor density greater than 1 tends to settle; an important consideration when dealing with potential or actual fires where flammable liquids and gases are involved, as most flammable liquid vapors have vapor densities of greater than one, such as gasoline at 1.4 and ethyl alcohol at 1.9, while some flammable gases have vapor densities lighter than air, such as hydrogen at 0.6 and acetylene at 0.9.

Vapor Dispersion The movement of vapor clouds in air due to turbulence, gravity spreading, and mixing.

Vapor Drift *see* DRIFT.

Vapor Pressure The property which determines how easily something evaporates; the lower the vapor pressure, the more easily a substance will evaporate. Also, the pressure exerted by the molecules of a liquid leaving the liquid and entering the void space of a closed container; measured in terms of psi. Increased temperatures applied to a closed container will cause the vapor pressure to increase. When the vapor pressure within a closed container exceeds the capability of the container to withstand internal pressure, a mechanical rupture of the container will occur. *see* BLEVE.

Vaporization The process of becoming a gas.

Vaporize To evaporate; to form a gas and disappear into the air.

Vaporizer A device for converting liquid to vapor by means other than atmospheric heat transfer.

Vaporizer, Direct Fired In LP-Gas, a vaporizer in which heat furnished by a flame is directly applied to some form of heat exchange surface in contact with the liquid LP-Gas to be vaporized.

Vaporizer, Indirect A vaporizer in which heat furnished by steam, hot water or other heating medium is applied to a vaporizing chamber or to tubing pipe, coils or other heat exchange surface containing the liquid LP-Gas to be vaporized; the heating of the medium used being at a point remote from the vaporizer.

Vaporizer, Waterbath (Immersion Type) A vaporizing chamber, tubing, pipe coils, or other heat exchange surface containing liquid LP-Gas to be vaporized, by immersion in a temperature controlled bath of water, water-glycol combination, or other heat transfer medium which is heated by an immersion heater not in contact with the LP-Gas heat exchange surface.

Vaporizing Burner A burner containing an integral vaporizer which receives LP-Gas in liquid form and which uses part of the heat generated by the burner to vaporize the liquid in the burner so it is burned as a vapor.

Variable Liquid Level Gage Device to indicate the liquid level in a container throughout a range of levels. *see* SLIT TUBE GAGE.

Variable Pitch Propeller An aircraft propeller with provision for changing the pitch setting while rotating.

Vectors Invertebrates which transmit infectious agents by biting through the skin or by depositing infective materials on the skin or on food; any animal or insect involved directly in the transmission of communicable disease, either to humans, animals, or to plant life.

Vent A device to control or limit tank pressure. Some types are: pressure relief; vacuum relief; fusible (opens at elevated temperature); frangible; Christmas tree (slang for a combination vent).

Ventilated Box Car (RR) Similar to an ordinary box car but suitable for the transportation of produce or other foodstuffs not needing refrigeration.

Ventilation (1) The physical release of smoke and gases from a structure which is on fire. (2) Method used to circulate heat, air conditioning and fresh air in a structure and to remove stale air or toxic fumes. (3) The act of getting oxygen to a patient's lungs through artificial respiration.

Venturi Tube Short tube with calibrated constriction to measure differential pressures in flowing gases or liquids.

Vermin Pests; usually rats, mice, or insects.

VFR Visual Flight Rules. Apply when visibility is greater than three miles and the flight path may be controlled by visual reference to the ground.

VHF (Very High Frequency) Radio spectrum between 30–300 MHz; includes EMS channels from 150–173 MHz.

Virgin Material Raw material that has never been processed in a manufacturing system, such as copper, aluminum, lead, zinc, iron or metal ores, sand, trees, or crude oil; usually uses more energy in manufacturing than recycled material (material which has been processed at least once).

Virus A disease-producing organism (pathogen) that needs living cells to grow and can cause disease in plants and animals including people; too small to be seen with a normal microscope.

Viscosity The thickness of liquids; the degree to which or the ease with which a liquid flows; usually increases when temperature decreases. A liquid with a low viscosity will flow very rapidly and any spill of that liquid will create problems very quickly; a high viscosity liquid will not flow as easily and can therefore be controlled more readily should it spill.

Visual Apporach An approach by an IFR flight when either all or part of the instrument approach is not completed and visual reference to terrain is used.

Volatile When a substance (usually a liquid) evaporates at ordinary temperatures if exposed to the air.

Volatility The tendency of a solid or liquid substance to pass into the vapor state; how quickly and easily a liquid or solid evaporates at ordinary temperatures when exposed to the air.

Volume Reduction The processing of waste materials to decrease the amount of space the materials occupy, usually by mechanical (crushing or shredding), thermal (incineration or pyrolysis), or biological (composting) processing.

Volumetric Filling or Loading Filling a container by determination of the volume of LP-Gas in the container; unless a container is filled by a fixed maximum liquid level gage, correction of the volume for liquid temperature is necessary.

Vomitus Stomach contents that are regurgitated; matter which is vomited.

Voltmeter An instrument used to measure the potential difference in an electrical circuit in volts.

Volumetric Efficiency The actual volume of fluid put out by a pump divided by the volume displaced by a piston or other device in the pump. Usually expressed as a percentage.

W

Waiting Period Time interval.

Wake Turbulence Disturbance in the air due to the immediate prior passage of an aircraft.

"Warning" Signal word used on pesticide labels to identify moderately toxic substance.

Wash To clean, usually with clear water.

Wash Over To release oil well pipe stuck in the hole. A rotary shoe, which cuts away the formation or mud, is made up on the bottom of the pipe and the assembly is lowered. Rotation of the assembly frees the stuck pipe.

Wash Pipe A short length of surface-hardened pipe, fitting inside the swivel and serving as a conduit for drilling fluid through the swivel.

Wash Tank A tank containing heated water through which crude-oil emulsion is forced to flow, used to remove the water from crude oil.

Waste Useless or discarded materials.

Waste Discharge Requirements An authorization by the Regional Water Quality Control Boards setting requirements for discharge of wastes from point sources.

Waste Exchange Waste clearinghouses where pretreated or untreated hazardous wastes are transferred; operating on the principle that "one man's waste can be another's feedstock."

Waste Management The total process of waste collection, from its point of generation through its transportation, treatment, and final acceptable disposal.

Waste Recycling Processes and procedures which allow industries to recover chemical resources for new or additional uses.

Waste Reduction Modifications in industrial processes which result in the production of fewer hazardous by-products; prevention of waste at its sources, either by redesigning of products or by changing societal patterns of production and consumption.

Waste Transfer Center A reception area used as an adjunct to a waste collection system; may be fixed or mobile.

Waste Treatment The application of physical, chemical, or biological processes to hazardous wastes which alter the properties of the waste materials and render them innocuous, reduce toxicity or substantially reduce the volume of material requiring special disposal. The most common technologies used to treat hazardous wastes are:

PHYSICAL TREATMENTS

DISTILLATION—Separating liquids with different boiling points by heating the mixture to vaporize and retrieve certain components.

FILTRATION—Separating liquids and solids by using various types of filters.

SEDIMENTATION—Removing suspended solid particles by providing time and space for settling in special tanks or holding ponds.

SOLAR EVAPORATION—Reducing liquid wastes in volume through simple evaporation in uncovered impoundment ponds.

CHEMICAL TREATMENTS

CHEMICAL OXIDATION/REDUCTION Detoxifying wastes by breaking the chemical bonds of a compound by the passage of electrons from one reactant to another.

NEUTRALIZATION—Reducing the acidity or alkalinity of waste by mixing acids and bases to produce a neutral solution.

PRECIPITATION—Removing compounds in a waste by adding a chemical to cause the formation of an insoluble precipitate.

STABILIZATION/SOLIDIFICATION—Transforming liquid wastes into a stable or solid form with high structural integrity, thereby rendering the hazardous components less mobile.

BIOLOGICAL TREATMENTS

ACTIVATED SLUDGE—Exposing wastes to biological sludge which acts on the hazardous components of the material (primarily used to treat waste waters).

LAND TREATMENT (LAND FARMING)—Allowing microorganisms in the soil to biodegrade organic wastes (primarily used for crude oil wastes and to be distinguished from "landfilling").

Waste Stabilization Ponds Allowing wastes to decompose over a long period of time.

Water Capacity The amount of water required to fill a liquid container full of water at 60°F (15.6°C), in either pounds or gallons.

Water in Oil Emulsion A solution formed when water is mixed with a relatively viscous oil by wave action and where droplets of water are dispersed throughout the oil; extremely stable and may persist for months after a spill. Opposite of Oil in Water Emulsion.

Water Pollution Contamination of water by unwanted material.

Water Reactive Materials Substances that react in varying degrees when mixed with water or when they come in contact with humid air; generally flammable solids.

Water Solubility The ability of a substance to mix with water.

Water Supply Officer Individual in the Command Post responsible for obtaining proper amounts of water or other extinguishing agents for the operation at hand.

Water Table The fluctuating upper level of the water-saturated zone (groundwater) located below the soil surface. *see* GROUNDWATER.

Water Tender (ICS) Any ground vehicle capable of transporting specified quantities of water.

Waybill Document prepared at the point of origin of a RR shipment showing the point of origin, destination, route, consignor, consignee, description of shipment and amount charged for the transportation service; generally forwarded with the shipment to the agent at destination.

Wax Any of a class of pliable substances of plant, animal, mineral or synthetic origin; generally consist of long-chain organic compounds and are included in the residue formed following the refining of crude oil.

Way Car Caboose.

Weather Interference (RR) Natural conditions which render loading or unloading a car impracticable.

Weathering The alteration of the physical and chemical properties of spilled oil through a series of natural processes which begin when the spill occurs and continue indefinitely while the oil remains in the environment.

Weed Any plant growing where it is not desired.

Weight Agreement A contract between shipper and carrier usually following a series of weighing tests, under which carrier agrees to accept shipper's goods at certain agreed weights.

Weight Filling Filling containers by weighing the LP-Gas in the container. No temperature determination or correction is required, since a unit of weight is a constant quantity regardless of temperature.

Weir Skimmer A type of skimmer which employs the force of gravity to drain oil from the water surface; basic components include a weir or dam, a holding tank, and an external pump. As oil on the water surface falls over the weir or is forced over by currents into the holding tank, it is continuously removed by the pump.

Well A driven, drilled, bored, or dug excavation, lower inclined from the vertical, with a depth greater than the largest surface dimension, generally of cylindrical form, and often walled by some means to prevent the excavation from caving in.

Well Car Depressed Center Flat Car.

Wetlands Areas where water is at, near or above the land surface long enough to be capable of supporting aquatic or hydrophytic vegetation and which have soils indicative of wet conditions.

Wettable Powder A finely ground pesticide dust which will mix with water to form a suspension for application; may not burn, but may release toxic fumes under fire conditions.

Wetting Agent An adjuvant which reduces surface tension and allows a pesticide to spread out and more evenly coat a surface; it decreases the rolling and running off of a pesticide.

Wheel Flange The projecting edge or rim on the circumference of a RR car wheel for keeping it on the rail.

Whole Body Counter Device used to identify and measure the radiation in the body of people and animals; uses heavy shielding to keep out background radiation and ultrasensitive scintillation detectors and electronic equipment.

Whole Body Expousre The exposure of the entire body to radiation, rather than an isolated part such as an arm, foot or head.

Wicking Agent Substances such as straw, wood chips, glass beads and treated silica which are used to increase oxygen availability and provide insulation between oil and water during the disposal of spilled oil by burning.

Wind Shear Change of wind velocity with distance along an axis at right angles to the wind direction (usually vertical or horizontal).

Wind Sock Cone-shaped cloth sock which hangs in an open area of the airport to serve as a weather vane, indicating wind direction.

Wireline Wire rope.

Wire Rope Rope used in the oil industry composed of steel wires twisted into strands that are in turn twisted around a central core of hemp or other fiber to create a rope of great strength and considerable flexibility.

Wood Preservatives Pesticides used to treat any wood to prevent insect damage, dry rot, or other damage.

Wood Pulp Primary material from which most papers are made; made of small, loose wood fibers mixed with water.

Working Pressure Maximum pressure at which an item is to be used at a specified temperature.

Work Over To perform one or more of a variety of remedial operations on a producing oil well to increase production.

Wrench Flat A flat area on an otherwise round fitting to which a wrench can be applied. A wrench square.

Wye Triangular arrangement of tracks on which locomotives, cars, and trains may be turned to reverse direction.

X

X The unnamed SI unit for measuring radiation exposure. 1 unit = 1 coulomb/kg = 3876 roentgens.

X-Ray A penetrating form of electromagnetic radiation emitted either when the inner orbital electrons of an excited atom return to their normal state (these are characteristic X-rays), or when a metal target is bombarded with high speed electrons (bremsstrahlung). X-rays are always nonnuclear in origin.

Y

Yard (RR) A system of tracks branching from a common lead or ladder track, with defined limits, used for switching, making up trains, or the storing of cars.

Yard Clerk (RR) Person engaged in clerical work in and around RR yards and terminal.

Yarding in Transit Unloading, storing, sorting, etc., of forest products in transit.

Yaw The rotational motion about a vertical axis.

Yield The total energy released in a nuclear explosion; usually expressed in equivalent tons of TNT (the quantity of TNT required to produce a corresponding amount of energy).

Yield Point Maximum stress a solid can withstand without undergoing permanent deformation. Tensile strength.

Z

Z The symbol for Atomic Number.

Zero Tolerance (Obsolete) Where, by law, no detectable amount of pesticide may remain on any agricultural commodities when offered for shipment. Extremely sensitive detection methods now available have forced change in law.

Zone As applied to oil reservoirs, describes an interval which has one or more distinguishing characteristics, such as lithology, porosity, saturation, etc.

Zone of Influence Maximum extent to which a waste disposal facility will affect surface and groundwater quality.